(Coach)

I Was Compost When Compost Wasn't Cool

My Forty Years of Trials, Tribulations, Failures, Successes, Mentors, and Memories in the Business of Composting and Farming

Stevan A. (Coach) Brockman

AuthorHouse™
1663 Liberty Drive
Bloomington, IN 47403
www.authorhouse.com
Phone: 1-800-839-8640

First published by AuthorHouse 8/11/2010

ISBN: 978-1-4520-4921-2 (e)
ISBN: 978-1-4520-4922-9 (sc)

Library of Congress Control Number: 2010909802

Printed in the United States of America
Bloomington, Indiana

This book is printed on acid-free paper.

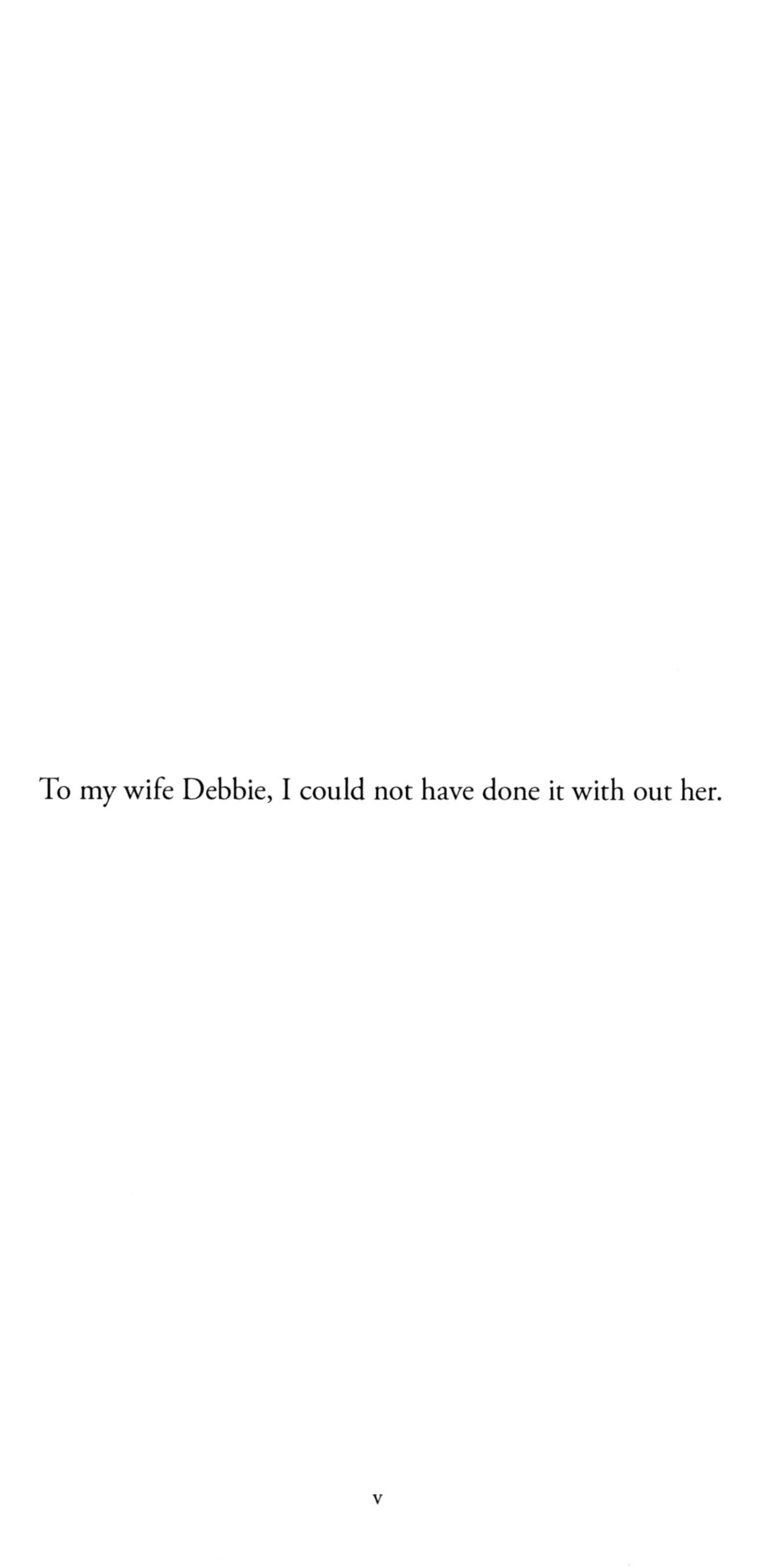

To my wife Debbie, I could not have done it with out her.

Contents

Forward 3
Introduction 5
The Early Years 13
Hay, Campbell Soup, And College 25
Family Growers And Campbell Soup 39
Death, Farming, And The Ladies 51
Compost Products; Birth And Death 61
Life After The Mushroom 77
Machinery, Innovation, And Imagination 97
More Machinery And More Memories 111
More People And More Problems 125
Stable Bedding And Customers 137
My Next Forty Years 145
Acknowledgments; My List Of Most Respected 153

Forward

As far back as I can remember, I loved the world of Farming. The first time I was able to grow something out of the earth, I became addicted. The older I got and the more diversified the farming operation became, I became addicted to the world of Composting. I had a double addiction that has remained very strong without interruption for more than four decades. The following pages will certainly give the reader some insight to my last forty years in the very seldom told story of Farming and Composting and on what most would consider a big time scale. You will get a glimpse of the interaction I had with People, Machinery, Materials, Money, and most important, Mother Nature. It has been both enjoyable and emotional to write about all the conflicts and conquest I was a part of for the last four decades.

You will be able to read about a young man's journey into the world of Farming with absolutely nothing and transcend into the biggest operations of its time. You will also be able to read about the world of Composting. It has not always been a glamorous topic of discussion. Everybody has gone "Green" and articles are being published everyday that the world is getting warmer. Thousands of articles about the cause, but little on actual cures. Composting is sure not the cure-all, but it can go a long way in cleaning up most of the problems as you will discover. There is no doubt, I was green before green was cool, and composting plays a big part.

There is plenty of opportunity to meet many of my mentors and role models that have guided me through these forty years. I hope your list is as long and as fortunate as mine has been. Unfortunately, many left this world far too soon and continually wonder what might have been if they were still here?

My addiction to Composting has lasted seven days a week for over forty years, stronger than either alcohol or drugs could ever have effected anyone. I was cut off from Composting cold turkey four days before Christmas in 2009 without the help of therapy. Fortunately, Farming is still an option to turn to if I need some type of psychological help to relieve my dependency from Composting. I hope it will be enough.

Most of the book is autobiographical because of a need to talk about my family, my upbringing, my work ethic, and my forty years with all the wonderful people that I have worked with. There is some science as well because the amazing story of Composting has to be told at the basic level. Economics comes into play throughout the book because it is a very big part of the Farming and Composting industry. There is also a lot of good old stories I hope you will enjoy and be able to share with others.

I have to devote this book to my wife, Debbie. She has been with me thirty five of those crazy forty years enduring all the good days as well as all the bad days. No way to ever thank her for all the support.

I hope you have as much enjoyment reading this book as I did remembering and writing, "I Was Compost When Compost Wasn't Cool."

Introduction

In the beginning, God created the Heaven and the Earth. Thank God for me, He also created Composting and Farming. Nobody knew what to call either for the first million years or so, but everybody knew it existed. The first time a leaf fell or a blade of grass died, the process of Composting started. If it didn't, all that dead material would have been over our heads sometime in the third century. Farming, the second oldest profession on earth, started because people got tired of eating dinosaur meat and had a taste for something else.

Composting began years ago as the decomposition of organic material to a stable or inert state. More recently, it has become the controlled process of decomposition for the almighty dollar. For those of us who do the controlling, it is the movement or handling of organic materials that take it from the raw form to the usable inert form. It all boils down to logistics, and the process has three phases. It starts with the collection and delivery of the raw feedstock to a site for the process to begin. The actual process of Composting takes place and then the use or sale of the finished product can occur. All this sounds rather simple until you interject manpower, machinery, management, money, and mother nature into the formula.

Many operations spend too much time and worry about the physical and chemical process that takes place. Remember, composting was created well before man and was doing quite well before any of us tried to control it and make money to do so. The biggest reason operations fail is for the lack of thought about all the movement involved with the entire process. As long as money is received in a timely manner for any part of the logistic trail, the operation will normally succeed. The more money received for any part of the process, the better the machinery, manpower, and site can become.

At the age of seventeen, I saw, smelled, and felt compost for the first time, and immediately became addicted. Not just by the look of compost, but the way it was transformed from a raw product into a useful material. Interested in all the ways compost could be used. What could and could not be composted into a useable product. The biological processes that were happening inside that pile of raw ingredients. The effects mother nature had on the process. And finally, the income that could be derived from composting on the inbound side as well as the outbound. It was amazing how all of this could intertwine with my farming addiction and work so seamlessly.

It was my good fortune to grow up in a rural, middle class environment five south of Joliet, Illinois. My father, mother, my two older brothers, my younger sister and I all pitched in whenever we could to help make a comfortable life in the early "fifties". My father worked shift work for the Texaco Oil Company in Lockport, Illinois about ten miles north of our home. He started there in July of 1929, about three months before the stock market crash. He worked until 1972 when he died of colon cancer. He was only fifty nine years old at the time.

He once told me the story of how he got hired in the year of the depression. He had gotten word that the man in charge of hiring at Texaco was having a party at the only tall building Joliet has ever had, The Woodruff Hotel. He went to the hotel and was told the party was on the top floor. He wanted to see him in the worst way trying to push his application through for hiring the following month. The man answered the door and appeared to be very intoxicated as far as Dad could tell. He only had one question for Dad and that was to be sure he was eighteen years old. Dad answered "yes" even though he would not be eighteen for another two months. He did get hired and the age problem on his record was not changed for about ten years, after the man that hired him had left the company.

Dad spent most of his working day outside in the weather. If overtime was offered, he was always the first to raise his hand. He never liked to work on the evening that the clocks were pushed back an hour in the Fall because they never got paid for it at the plant. He hoped to work the following Spring when the clocks were pushed forward and get that hour back. I don't know how he worked like he did all those years with one week working day shift, the next week working afternoon shift, and the third week working the midnight shift. He did this for over forty years without ever complaining as far as I can remember. I think most companies now outlaw such practices because of the detrimental effects it has on human life.

Dad was never one to sit around. He drove the local school bus for over thirty years when he had time off. He drove a fuel oil delivery truck in the winter and raised a huge garden in the summer. He started a roadside market in front of the house trying to lure customers off the busy road that ran by the property. If dad was working, us kids would go out in the garden and pick the produce that was ready and set up the stand for the day. If we were already done selling for the day by the time dad got home at four thirty, he could tell it was a good day selling vegetables. I guess it was the first "Farmers Market" that are so popular today that he started in the late fifties.

He was an excellent provider to mom and us four kids in tough times. He only missed work one day in his forty years of service to Texaco. A very freak snowstorm came in very late one evening, and dad was supposed to go to work the next day. All the roads were closed because of two feet of snow and gale force winds. The two most unusual aspects about the storm was that no weather forecast had been made for such a storm, and it was in the first week of April of 1957. Whenever Dad knew of bad weather, he had friends in Lockport he could stay with and actually walk to work if he had to. I definitely took after Dad in many ways. No matter how sick, I never missed work.

Mom was a homemaker in every way possible. She raised us four kids, took excellent care of the house, did all the laundry, got all of us to church on Sunday even if she had to get up early and take Dad to work so she could have the only car for the day. In the summer, she and us kids did a lot of work for the local Greek truck farmers that dotted the area where we grew up. We did a lot of onion peeling, cleaning beets, and Swiss chard bunching in our old two car garage after supper. All of us kids would pitch in as mom would say, "we were earning money to pay the taxes on the property," when it was actually for Christmas gifts. It was not uncommon to peel fifteen to twenty thousand onions in a good evening.

The onions would come to us fresh out of the field in wooden crates with all the dirty skin still attached. Peeling was a process that took all this dirty skin off with our fingers and reveal that beautiful white onion underneath. Us kids would do that, and after thirty or so, hand them over to mom for final inspection and bunching. She would, in an orderly fashion, put eight or ten onions into a bunch and put a rubber band around their neck. We would receive one and a half cents for each bunch delivered. It was not uncommon to make fifteen or twenty dollars on a good night out in that old garage and a long way to a good Christmas. The only problem I can remember, beside worn out fingers, was the night a large skunk got inside the garage just before we were going to start bunching some Swiss chard. Luckily, dad was home and had a twenty-two rifle that never did kill the thing, it only persuaded the skunk to leave so we could go to work. It did not let out its normal stink either, or the resale value of all that Swiss chard would have been nil.

Dad grew up about three hundred feet from where he would raise his own family. The house we lived in was actually an old converted Texaco Gas Station with a drive through canopy, a large dinning room, and a large kitchen where my Grandmother would cook meals for travelers. This building actually sat on the old Route 66 roadway that is so famous going from Chicago to LA. All the gas was pumped by hand up into a ten gallon glass container and then delivered to the car's gas tank by gravity. I sure wish I had that pump now because they are very valuable. I never knew why the container was made of glass until about ten years ago. It seems the motorist wanted to see what they were buying because there was a lot of dirty gas being sold back in those days. After my Grandfather died in 1939, they stopped selling gas. Dad moved the old building three hundred feet north, remodeled it into a home, and when Mom and Dad got married, they moved in to start their family in 1940.

Mom, on the other hand, grew up over two hundred miles south of the Joliet in a very remote part of Illinois. The farm sat about a mile from the Wabash River near the small town of West Union, Illinois. She lived out on what they called the "prairie" where my Grandfather worked and farmed the land for the same absentee landowner for over fifty years. Many summers at a very young age, I would stay with my Grandmother and Grandfather for a few weeks at a time and is probably where I got my initial addiction to farming. They were always showing me all the aspects of old time farming like milking cows by hand, working in the garden, feeding the chickens and hogs, putting up hay for the livestock, and getting up before the sun everyday.

My Grandfather never went to church, but he never worked on Sunday either. He took my Grandmother to church each and every week without fail. They got married when they were both seventeen with only seven dollars between the two of them. They started a family with my Grandfather only receiving one dollar per day for his labors in 1915. They celebrated fifty years of marriage and the loss of my Grandmother in 1965. My Grandfather continued to farm until he was about seventy five, moved off the farm and into the old town of West Union where he died in 1983 at the age of eighty six.

For some unknown reason the three kids, including my mother, all migrated up into the Chicago area after they graduated from high school. My Mothers high school was located about twelve miles south of the farm. She drove the old Chevy car to and from school starting at the age of fourteen. One of her greatest prides was that she took and passed Latin all four years while attending Hutsonville High School. All three left that old farming community behind for greener pastures. My Great Aunt Ty, lived in Chicago and fixed all the kids up with their initial dates and eventual spouses.

My grandfather and Grandmother never needed anything except flour, sugar, and coffee from the grocery store. They would ride into the little town of West Union every two or three weeks and trade eggs, butter, or meat for the three things they needed. No money ever changed hands. They never realized there was a depression since they raised everything else they needed to survive out on that prairie. My Mother and her two brothers were always visiting the folks at least four or five times per year after they got married and had their own children. They were forever orchestrating the visits so not all would descend on the farm at the same time coming out of the Chicago rat race and back down to the prairie.

There was a lot of canning and freezing going on year round at the farm. My Grandfather's garden raised enough to feed not only themselves, but all of his kids and eventual grandkids in the process. There was always a huge pile of potatoes in the root cellar and it was my job before going back home to put enough in a sack to get us by for several weeks. There was hundreds of jars of fruit and vegetables canned down in that cellar, and nothing tasted any better than some canned peaches on Christmas day out of one of those old Ball Jars. Everything tasted better when my Grandmother fixed it. One of the granddaughter's married a guy that had never been to a farm or eaten a farm fresh egg the way my Grandmother could make them. She told the story that there were fourteen eggs in the house when he started to eat one morning on his initial visit to the farm. After he went through those, my Grandmother directed him to the hen house to get more. He brought back an additional half dozen, and polished them off as well. They would always laugh whenever anybody brought that story up on the amount of eggs that kid could eat.

That prairie land that my Grandfather farmed all those years was one of the most productive areas in the U.S. The biggest problem was the Wabash River. It might not let you plant in the spring because of flooding. You might have everything planted and the river would flood everything out, and you must replant. You might be a month from harvest, and everything gets wiped out. Some years everything goes just right, and the crop is unbelievable. The land was referred to as the "bottoms" because it laid so close and low to the river. Some of the most challenging land I have ever seen and nothing could be done to improve it. One very tough year, Grandpa had the corn "laid by" before Mothers Day. That meant he had it planted and was tall enough to be cultivated by the second Sunday in May. The river came up and drowned everything out. He had a second crop planted and "laid by" a second time by Father's day. It was some amazing property.

Another very distinctive story from out on the prairie was the Gypsies that used to stop by the farm for free handouts. They used to scare my Grandmother to death when this band of travelers would visit, usually when my Grandfather was out in the field. She would give them anything they asked for hoping they would quickly leave. One time my Grandfather was close enough to the house to see a band of Gypsies pull into the yard. He stopped the old tractor, walked up to the house, and told the group to wait one minute because he had something for them. He came out of the house with a ten gauge, double barrel shot gun. He asked the group to leave, and my Grandmother never had a problem with Gypsies after that day. They must have marked something along the road so other caravans would not attempt to stop for a free handout.

I only knew my Dad's mom for a short time. She died in 1957 when I was only six, and his father and one of his sister's died in 1939. His family of four sisters and one brother were always visiting our house out in the country to get out of Joliet for all the occasions. They and my Mom's family were all very close and very similar to my two older brothers and my younger sister today.

The Early Years

My two brothers and I all started out at about age seven or eight working for the local Greek truck farmers. There was about ten or twelve families around the Joliet area that produced a lot of fresh vegetables in the summer and "trucked" them to the Water Street Market in Chicago for sale. There were also a lot of summer jobs for kids that wanted to work. We would weed onions, pick green and yellow beans, pick lettuce, endive, asparagus, kohlrabi, beets, turnips, and tomatoes most of the time on our hands and knees. We would reap a solid thirty five to forty cents per hour. Not bad for the late fifties and I think this is where the government got the idea for minimum wage in this country.

My oldest brother got a job with Caterpillar in Joliet after he finished high school. Not long after he got hired, he was drafted by the army for a two year tour of duty that landed him mainly in Germany. When he came back, he went straight back to his same job at Caterpillar that he had left two years earlier. He also got involved with restoring old Chevy cars and has a collection of vintage automobiles who's value is doing much better than my stock market portfolio. He retired when he was only fifty two and he and his wife enjoy traveling and working with all his old cars.

My middle brother worked for the local ammunition plant making bombs after high school. After a few years there, he enlisted in the army for three years stationed in France for most of the time. After he came home, he got married and settled about twenty miles south of Joliet working at about four or five different jobs in the area. When the government decided to build the Abraham Lincoln National Cemetery near Elwood, Illinois, my brother was the first to be hired and the first to be fired.

At this cemetery, when a single body is buried, the hole is dug about six foot deep. After digging, the operator jumps into the hole to square the corners up so the coffin will fit just right. When there is going to be two buried in the same hole, the operator digs down nine feet, places a protective devise around the hole so the dirt walls will not cave in on him or her when squaring the corners up. He was digging one of these deep holes and the boss came by and said, "there was not enough time to get and use one of the protective devises". My brother would not enter the hole without it, and got fired. He tried to appeal his dismissal to a local politician, but never got any satisfaction as the politician himself was charged with improprieties in Washington about the same time. My brother was fortunate to find another job, and moved on.

My sister, beside helping mom with the all the house chores, got a job near the high school when she was just a freshman. Four months after high school she got married, and lives only about a half mile from where she grew up.

She and I were only thirteen months apart in age. We did everything together most of are early years. One of the worst days occurred when she was about five years old. We were out near the garden and I was chopping down old cornstalks with of all things, an axe. For some unknown reason, she bent down in front of the path of the swinging axe and got struck in the head. I was mortified with her crying and the blood coming out of the cut on her head. We both ran into the house with a trail of blood left behind all the way from the garden and up the back steps. Mom calmly placed a cold wet rag over the cut, the bleeding stopped, and we all considered her very lucky.

The older I got, the more my physical labors drifted to local farmers who needed help baling and stacking hay, shelling corn out of corn cribs, and walking soybean fields to pull weeds and volunteer corn out between the rows. In the winter, I would hitchhike into Joliet and shovel driveways and sidewalks, and in the summer I would mow grass for some extra money. Some of the hardest work was picking sweet corn by hand early in the morning when all the dew was still on the plants. You would get soaking wet from the armpits down and no way to dry off. In 1958, I bought my first new bike with money earned picking green beans. I washed and waxed it almost everyday. I did make one mistake by riding it to my friends farm that had goats. The bike came with one foot long brightly colored streamers coming out of the handle bars that the goats found very tasty.

I might have had a little glimpse of compost at that early age by seeing some old rotten hay in the corner of the barn or by noticing that the pile of grass on the property line never got any bigger no matter how much grass I cut and piled there. I did realize that both produced an odor when they were disturbed.

I attended the local high school, but because of rheumatic fever in the winter of my eight grade and a wonderful agriculture teacher, I was able to leave school around one o'clock each day if I wanted for work related purposes. The school would not let me take physical education because of insurance reasons connected with the fever. That combined with study hall at the end of the day, meant I could leave after one pm if I wanted to.

Ron Deininger, my high school Agriculture teacher, would become a very important part of my life but did not realize it as a freshman. He was about fifteen years older than me. He never stopped working for as long as I have known him. He taught high school agriculture related classes during the day, drivers education during the afternoon and weekends, raised hogs and farmed at night and early morning before class. If chores for the hogs were part of the routine for that particular morning, hog manure could sometimes be smelled in the classroom. He was also our "FFA",

Future Farmers of America, advisor for all four years. He pushed all the kids he had contact with to our limits each day. The biggest regret I had was not buying land as he suggested to do all the time. Over the years, he and his son and daughter have become very successful in their purchase of property.

One of the biggest highlight while in high school was being a part of the FFA. The high school was set in a rural atmosphere at the time I was attending. There were a lot of students from the rural areas around the east side of Joliet probably totaling about fifteen percent of the total enrolment. There was a lot of agriculture related classes devoted to these selected students and I tried to take all that I could. Most of the friends were from farm backgrounds where I was just from the country but enjoyed farm life. The four years of FFA under the supervision of Mr. Deininger were very enjoyable and educational. He entered us into every type of competition that was ever conceived. We had judging teams for grains, weeds, meat, poultry, dairy and beef cattle, soils, extemporaneous speaking and about a dozen other things that we competed for in our area and at the state level. He made time to drive us personally to either Champaign or Springfield, Illinois for the state judging contests on many occasions. During my four years and for a long time after I left high school, the Joliet Chapter of FFA was known through out the state as the one's to beat at any competition all because of Mr. Deininger. In my senior year I was picked as a Star State Farmer from the state of Illinois. I was up against guys who had large grain and livestock operations. I won because of my meticulous record keeping from all the farmers I worked for as a young kid and all the income earned from jobs I had done. Only one percent in any given year receive such an honor and was all because of Ron Deininger and his support.

When I did leave early from school, Mom would pick me up if she had the only car we had to avoid the six mile walk. When she did not have it, I would start walking and hitch hike home usually getting a ride in less than a minute. I would quickly get changed, and hitch hike another two or three miles to help some farmer in the area. It is sure amazing how well hitch hiking worked back then compared to now.

In the summer of 1965, I met Gorman O'Reilly from the eastern part of Joliet. He would eventually become my first full time employer. I owe everything I would become because of his tutelage over me the next seventeen years of my life. He was a very unique individual with both a strong mind and a strong back. His farm enabled me to work after class, on the weekends, and all summer long. He would even come and pick me up from school if he had time and take me home at night so I could do my homework. While I was in college, I would work any weekend I came home, and all summer long. The day I graduated from college, I started working fulltime for Gorman.

Gorman was the lone boy from a family of six girls. He was smart enough to finish high school by the time he was only sixteen, and asked his father if he could sign off for him to join the Navy by age seventeen. He did not see action but did get on a destroyer for a while during his tenure in the Navy. When he was discharged, he worked at a lot of hard labor type jobs and finally settled into farming in the early fifties. He settled on the home base of operations in the early sixties and is where I met him.

I have always worked from a young age starting at eight or nine with the local truck farmers, by age fourteen I started working for grain and livestock farmers in the area, and by sixteen for Gorman. I have always enjoyed being out in the fields as far back as when I went to my Grandparents farm. I was a little big for my age as far back as I can remember. In eight grade, I contracted rheumatic fever close to my birthday in January. I endured a temperature of 107 for about three days and the doctors felt that I probably would have some heart damage. My two week stay in the hospital included a steady diet of Jello, morning, noon, and night because my throat was so inflamed. My late night snack would also be Jello. It took the better part of thirty years to enjoy Jello in my diet once again. When it came time to help local farmers with all the labor involved with Spring planting, I went right back after it with no side effects, but against my mother's wishes.

Having to go through a two week hospital stay was probably the low point of my year in eight grade. The high point had to be that Jesse Owens, one of my role models and a person that I studied and looked up to, was the commencement speaker for our graduation ceremony. Don't ask me how he was talked into coming to Joliet and speak to a bunch of eight graders, a class of less than sixty, but there he was and I was able to get his autograph on my program. I have always cherished that night and that autograph. Come to think of it, I only have two autographs in my possession. The other is of Pete Rose on a baseball that cost me twenty dollars and about an hour of my time. Despite his problems with gambling, I still think he was one of the best to ever play the game.

I have also enjoyed football ever since I can remember. Us kids always had a football game going in the front yard with some of the neighbors. My grade school did not offer football, but it did have basketball, baseball, and track that I did participate in and receive all area honors. In the late summer of 1965, I tried out for the freshman football team at my new high school and liked the running back position. It was a very common practice to line up all the varsity lineman and have the little freshman try and run through this wall of humanity on the first day of practice. My first pass through and all sixteen varsity lineman were laid out on their backs wondering what had hit them. For me, a little big for my age, it was nothing compared to baling and stacking hay all day. All the coach's saw what had just happened and immediately wanted me to go from the freshman team to the varsity team overnight. I went home that night hardly able to get my head in the house. The next day, I received a varsity outfit and was all set to go at it again until one of the coach's came up to me to inform me of non-insurability because of the rheumatic fever I had nine months earlier. I could not play football, I could not take gym, so I scheduled my classes so I could leave the school at one in the afternoon and go home to do heavy physical labor on some farm. My football career was over.

There was no extra money for me to attend college, but because of that wonderful agriculture teacher from high school, a full ride, all expenses were covered by a seldom used scholarship that he was able to uncover for me. The scholarship came out of the Division of Vocational Rehabilitation from the state of Illinois. I had developed severe pain in the arch of both feet at about the age of twelve. Since dad had very good insurance, mom took me to a foot doctor that explained that my foot grew a lot faster than the tendons holding the arch together causing the bad pain near my heal. He suggested a series of steroid injections right where it hurts. The pain from those injections was far worse than the pain from the actual problem. He stuck me ten times in each foot over a two year period with little relief. I can still feel those injections in my sleep some forty years later and wished I had never done that. I had mentioned to Mr. Deininger that I had under went this treatment and he must have put it in his mind somewhere and brought it out when it was needed. A four year ride, all expenses, and all I had to do was maintain a 3.5 grade average and get myself to the school. There is never a way to thank him for this and a lot more.

During my high school and college days, Gorman farmed a lot of U.S. government owned land that was known as the Arsenal. This area that encompassed more than forty thousand acres near Elwood, Illinois was taken, literally, from the landowners back in 1940 and 1941 so a military reservation could be built to produce bombs and ammunition for World War Two. There were 420 farm families uprooted and moved off the property for a price of less than three hundred dollars per acre, no more and no less. Everybody got the same price no matter where you lived on this forty thousand acres. Some of the land was the most productive in Will County, some was the least productive. All got the same price.

My wife's uncle tells the story that his father and uncle, two of the largest landowners in this whole uprising, fought the government for two years and ten thousand dollars in court fees only to receive their walking papers and the same price as everybody else. Her uncle is ninety five and still gets furious over what the federal government can do to a person if they want to. A lot of people wonder why this site was picked. Some thought of the good rail system in place. Some thought there was good water and road transportation. My wife's same uncle believes the real story is that the federal government looked at forty thousand acre parcels around the greater Chicago area and discovered this particular parcel had more money loaned against it than any other they could find. The government thought these farmers would fold very easily because of the financial burden and they were right. Everything was bought in less than two years.

All of this land was located about ten miles south of Gorman's headquarters near the town of Elwood, Illinois. Anything the government deemed excess and was not torn up with railroad tracks or buildings was leased back to farmers to farm on a five year cash rent basis. At first, so many farmers were mad at the government that a lot of the land was not farmed at all. Parcels were rented out for next to nothing and government officials actually went out into the area trying to entice farmers to take advantage of the situation. Gorman started to farm some of this land back in the late fifties.

There was one big stipulation to farming this land. You could not grow any government supported crops like corn, soybeans, or wheat. What was allowed was popcorn, hay , vegetables, or the pasture of animals. When I started to work for Gorman, he had a good deal of each. His hay acreage was devoted to mainly Alfalfa. Most farmers shied away from this because of the intensive amount of labor involved, but not Gorman. He possessed both a strong mind and a strong back. A lot of his production went to horse and dairy farms in the area and for his own use with his herd of beef cows. His popcorn production went to Cracker Jack and Jolly Time on a contract basis. More about the hay and popcorn later in the book.

When I started working for Gorman in 1965 on a more regular basis, he had a very large beef cow herd using up a lot of pasture ground rented from the government. Early in the seventies, a lot of the pasture started to get plowed up for crop production because it brought in more money than if it was used for a pasture. More crops like popcorn, rye, broom grass, and alfalfa hay started to be grown. Gorman's cow calf operation could not afford to pay for the decreasing amount of pasture and the increase in price farmers were paying. He made a decision to sell the entire herd, some 400 cows, to one buyer.

I will never forget the day the buyer showed up with all his trucks to haul all of the cattle off the Arsenal property. This particular pasture was about three miles long and on that very hot day, most of the cattle were located at the furthest point from the loading dock near a big grove of trees. The buyer drove a couple of pickups down to the end of the pasture and started a real stampede. The animals were crapping and peeing the whole three miles back to the trucks. The new owner saw dollars coming off all the cows as he was buying them over the scale per pound. It was well over a hundred degrees that day and I thought we would lose some animals let alone all that weight. About two hundred yards before the loading shoot stood an acre pond of fresh water. All those thirsty animals ran directly into the lake and each drank well over a hundred pounds of water before they could be loaded into all those trucks. All that extra cash into Gorman's pocket and out of the new owners. What a way to end your career with the cow herd.

In my seventeen years with Gorman, we did some very amazing things. The accomplishments were numerous and you will be able to read about them in the following chapters. I do want to describe two near death experiences I had as a newlywed working for Gorman that I never told anyone, not even my wife. I hope she will forgive me when she learns about the incidents. The first involved electricity, the second involved water.

It was a very hot and humid day, and I was cleaning out a grain bin after most of the corn had already been emptied. There are small auger type devices that pull the left over grain from the outside to the center of the bin so it can be loaded out with an auger that runs underneath the floor and out to a waiting truck. The big advantage with these little augers is it requires very little shoveling. These little augers, known as sweep augers, are run with a five horse electric motor and a set of belts. A couple of days before I used the auger, one of the employees had damaged the cord, and rewired the motors connections backwards so the hot wire was now in the place where the ground wire usually goes. Unbeknownst to me or anyone else, when my helper turned the switch on, I was electrocuted to the point of begin knocked down inside the grain bin. I laid on the bin floor for about ten minutes before I knew what happened. It didn't help that I was soaking wet from being over one hundred degrees inside this metal bin. The thirty amp breaker fortunately tripped before it killed me. The single strand of wire that I used to hold onto the auger while it did its work burnt a thin line in both of my gloves and left a distinct line in both of my hands that you can still see some thirty-five years later. Whenever I am around electric now I always double check everything before starting. I told the man helping me never to tell anyone about what happened, including Gorman. Needless to say, the wires were switched back the correct way before anyone else got hurt.

I also had a near death event with water. In early Spring of 1976, I noticed that beaver had backed up about two miles of water in a major creek in the Arsenal. They had put a dam in front of a six foot diameter culvert that ran under three sets of railroad tracks. There was over three thousand acres affected and water was covering a lot of my land in some places to the depth of five feet. I contacted Gorman on the radio and informed him of the problem. He told me to see if there was anything I could do to get the water to flow, but to be very careful.

Initially, I attempted to try and pull some of the long branches away from the mouth of the large culvert with very little success. The pressure against the mass of mud and limbs was unbelievable. I then had the very stupid idea to walk inside the culvert from the discharge end and see if I could pull a few logs out of the way from the dam to get some water to flow. It was almost pitch black inside this eighty foot long pipe, but I could hear a small trickle of water coming through this mess of branches and limbs. I got a hold of one good size limb and pulled as hard as I could. It was like somebody had flushed the culvert like a toilet sending me out the other end like a big wet turd. The wall of water rolled me over about ten times and down the stream about a hundred yards before I could get my arms around a tree along the bank and pull myself out of the icy cold creek. I was soaking wet from top to bottom, shivering, but never happier to be alive.

The beaver dam was removed, but I never told Gorman or my wife what an idiot I was that particular day. I ran the heater on the old pick-up truck for about an hour to dry off and warm up. That's when I started to feel all the pain inflicted on my back and knees and was hard to explain to my wife where all the black and blue marks came from. I understand that a culvert of that size and with that much pressure behind it, could easily discharge over ten thousand gallons of water per second! Every time I drive by that old creek I still can't believe what I did and considered myself very lucky to be alive.

Hay, Campbell Soup, And College

The main reason Gorman wanted me as much as possible was because of his alfalfa hay operation. It required a lot of heavy manual labor and with my size, I was an excellent hire. He was known in the northern part of Illinois as the "Hay Man". Most farmers in the area did put up some hay, but nothing like the quantities he put up every year. I quickly learned that the biggest problems with alfalfa hay production is trying to sell something that has had a little rain on it before it was baled or is a little older than it should have been when it was cut. In 1968, Gorman was approached by a hay buyer, Arden Rathmon, from the Campbell Soup Company. What in the world would a soup company want with hay? They wanted to know if we were willing to sell them some hay for their mushroom compost operation in West Chicago, Illinois about forty miles from Gorman's home base? At first, Gorman was very reluctant to sell any hay because of all the whore stories that came from selling tomatoes from the Arsenal property to the Campbell Soup plant on California Avenue in Chicago. Many producers tried their hand at growing tomatoes only to be told they were not needed once they became ripe, or slow unloading of trucks that were accepted at the plant, sometimes sitting for days in line to get a load dumped.

The mushroom farm was located about forty miles due north of Joliet. It was built in 1947 and employed over 275 Spanish speaking people at the peak. The farm became known as Prince Crossing and sat on over four hundred acres of very valuable land right in the middle of West Chicago. From September to May, the farm produced mushroom compost, filled mushroom houses, grew and harvested mushrooms, and packaged mushrooms for the soup plants around the Midwest. The houses would sit empty for the summer months because it was too difficult and too expensive to try a keep the rooms cool enough to grow mushrooms. The labor force would then occupy the four hundred acres of land and grow vegetables for the soup plants and the Water Street Market in Chicago. There was also an area that contained very small houses for the labor force to live in on a year round basis.

Not only was the farm looking to buy hay, alfalfa hay in particular, but the more rain or the older it was, the better they liked it. We now would have an outlet for the biggest problems with alfalfa hay production. If we cut it at the right time and we baled it in good condition, we would sell it to the horse or dairy farmers. If it got rained on or was a little old, off to Campbell's it could go. This was a very early example of how Gorman was always thinking. A third generation grain elevator operator, Jack Tyler, once told me that, "Gorman was the only farmer he knew that made money with hired help." It was the way he could think things out and I believe a lot of that rubbed off on me, it had to, we worked side by side for over seventeen years

We were invited up to Prince Crossing after Gorman agreed to sell them some hay to see what they were doing with all this material. I believe this is where the initial addiction for composting came from and has stayed with me for the last forty years. I already had the farming addiction since I was a young kid down on my Grandfather's farm and working on my hands and knees for all those Greek truck gardeners in the fifties.

There was one area where corn cobs were being watered very heavily with a fire hose trying to make that beautiful reddish cob produce heat and start to turn black. There was five or six piles all in different stages of color and moisture with the oldest producing a large plum of steam from the heat within the pile when the loader touched them and rolled them over for the necessary oxygenation.

There was fresh stable bedding being unloaded from the local Chicago racetracks that included all the moisture and droppings from the horses, all the beer cans and bottles, all the twines from the hay and straw bales, and almost every other imaginable piece of garbage you could think of. Men tried to pull out as much as possible before the material was watered down and piled, but were only about ten percent successful at any given time. Garbage has always been and will always be the biggest problem with compost.

Then came our hay portion. The truck driver would cut all the twines off each individual bale and deposit them off the side of the trailer for a waiting loader to push them up into piles while a man watered the hay down with a fire hose. What a fascinating place with men and machinery going in about five different directions all concluding with a composted material for mushrooms to grow on for the next ninety days.

They had been doing this job for over twenty years at this site about five hundred yards from the actual mushroom houses. Clean stable bedding was becoming harder to get and more expensive, and there was a push to try a different formula. The three materials were blended using a mixture of ten percent corn cobs, forty percent hay, and fifty percent stable bedding. Chicken manure, cottonseed meal, urea, potash, and gypsum were added to the blend, mixed with a front end loader, and piled for a few days to let the composting process start. Once the material started to show signs of heating up, it would be picked up with the loader and mechanically placed into straight sided rows or "ricks" that were eight feet wide, seven feet tall, and about two hundred feet long with a machine called a "choke feed turner".

This special built machine resembled a large manure spreader. You would feed the material into the front with a loader, it would pull the material upwards with large chains to a rotating drum that would vigorously eject the material out the back forming these long loaf like ricks behind the machine. A man was required to stand by the turner at all times so he could manually move the machine forward every two or three minutes and keep it straight. Once these ricks were formed, the material starts the composting cycle producing tremendous heat and odor. Oxygen is pulled into the sides and naturally exists out the top of the rick like a chimney producing gases like nitrous oxides, hydrogen sulfide, and methane. Over the two weeks it normally requires for the material to become compost, temperatures inside the rick will exceed one hundred sixty degrees. Since the microbes need a fresh supply of oxygen constantly, the material was picked back up and placed back through the choke feed turner every two or three days.

Back then, these formulas were all top secret within the mushroom industry. All mushroom growth depends on a composted material and no one has ever successfully eliminated this phase. Mushrooms are fungi and have no chlorophyll to produce their own food. The compost is prepared to supply that food to the mushroom. The basic parameter that must be met is to compost the raw feedstock enough to supply food needed for a good crop of mushrooms, but at the same time have no nutrition available to supply other fungi or competitors to compete against what you are trying to grow. The microbe that perform this miracle, the Thermophilic Bacteria, thrive as long as there is water in the right amount, available carbon, the right amount of nitrogen, and the most important ingredient, oxygen. When these microbes multiply, heat is produced, the material turns black, the raw ingredients become very soft and moist, and now ready for the fungi to get nourished. Turning the material adds the much needed oxygen and water that can be added very evenly to maintain the moisture of the material at seventy two percent through the two week process. The volume of material will shrink by at least fifty percent and the ammonia levels will become very high. To grow mushrooms, all this ammonia has to be cleared out of the material and this is accomplished in "phase two" of the mushroom cycle.

Phase two starts by putting this special material into specially designed houses that are about twenty feet wide, about one hundred feet long, and fifteen feet tall. There are beds built the length of the house made out of hardwood, one row on each side spaced about two feet apart and stacked six beds high. The material is mechanically brought into the house and dumped on the five foot wide beds where men physically spread the compost out evenly to a depth of about twelve inches. This is an unbelievably dirty and very hot job and in the warm months of the year, it can not be described with words. I like to watch a show on TV that shows dirty jobs. There has never been a job that is as dirty or hot as this one and should be considered for one of his next shows in the future. Once filled with about 120 tons of compost, the house is then sealed up and the temperature of the compost rises quickly to 145 degrees where it is held and controlled with fans and fresh air for about four days. This helps to eliminate the levels of ammonia in the material and out of the house. The temperature of the material is then lowered about five degrees per day for the next eight days, and the phase two or pasteurization phase is completed.

The house is now ready to be planted with rye seed that has been inoculated with mushroom spores or what is called "spawn". Spawn is made in very intense facilities where the enviroment is extremely clean and no chance of contamination could occur that would effect the outcome of what type of mushroom was to be grown. Campbell's had their own special strains of mushroom spores that they raised in three separate facilities across the U.S. Mushroom production has now started and there are a thousand things that must happen over the next sixty days to get a good crop of mushrooms.

Our hay sales to Campbell's lasted about four years. They needed hay at a rate of forty tons per day on a six day per week sechedule. We supplied them with hay as often as we could over the first two or three months. Then, Campbell's decided that grinding or shredding the hay would probably be more advantageous to their compost operation. By making the hay a little shorter, the material would compost faster and at the same speed as the stable bedding. They asked Gorman to perform this service for them, and so he purchased a "WHO" tub grinder from Greely, Colorado for about $57,000. All of our hay plus all the other vendors hay would now be delivered to this tub grinder first, ground in the WHO, and then trucked to Prince Crossing for them to compost. We always had about a two week pile of material ground in case we had some mechanical problems, and about a two month supply at Prince Crossing. We hired a trucking firm that took two loads of this ground hay up to the farm everyday. The trailer was specially designed with a large chain and cross bars on the floor that hydraulically unloaded the material off the trailer. It was one of the first "live floor" trailers I had ever seen and worked extremely well.

The truck driver for this company seemed to work about twenty hours per day. Someone would be at the Arsenal where the hay was ground at five in the morning to load this driver with his first load. Someone would go back to the site at eight thirty to load him the second time. You could almost set your watch by him. When he was finished with his two loads to West Chicago, he would drop that trailer and run to St. Louis with another trailer loaded with something else. He hauled like this, six days per week without a miss except for one rainy winter morning.

It was getting close to when the driver should have been back, and we were starting to wonder where he was. It had rained most of the night and all morning, and we were wanting to go home and call it a day. He used to have a nickname for his truck that he babied all those years. It was called "Little Hirran" for some reason, we never asked. There were no cell phones back in those days, so we left the grinding site to go look for the driver, and spotted "Little Hirran" about a mile away with its hood up. He had blown the engine and our day suddenly got a lot longer.

This particular tub grinder was a machine developed in cattle country. With the help of rotating hammers inside a large drum, it would take hay and make powder out of it for easy consumption by the cattle. We were advised by the company that made this machine, if we removed some of the ninety six hammers, through experimentation, the material would come out to Campbell's desired length and would entice them to pay for this added operation. In the four years we used the grinder, we never pleased Campbell's with the length they wanted a single time, but we could put a ton of hay through the machine every minute without a problem and they paid good money to do so. It was an unbelievable piece of equipment.

Another machine was starting to be demonstrated out west that really reduced the amount of back breaking work involved with hay making. Gorman became very interested in this machine called a "Round Baler". He figured if it worked as good as it sounded, he could rent more Arsenal land cheaply, and raise more alfalfa hay, and sell more to Campbell Soup. Gorman had such a well known reputation for making hay in the northern part of Illinois that it almost created a bad situation on a land bid for hay ground at the Fermi Lab in Batavia, Illinois.

There was over three thousand acres of land in and around the collider facility going up for public bid. Gorman decided to drive up to see the site and the bidding process. When he entered the room, he noticed that several farmers left the room only to return about ten minutes later. The bids were opened and read out loud publically but Gorman's name was never mentioned. He felt it was too far away from his home base and did not bid. Several of the farmers came up to him afterwards in disgust saying when they saw him walk in, they went out and raised their bids substantially. Fortunately, no punches were thrown and Gorman never went back to that area again. The men that did receive the top bid, sold most of that hay to Campbell's, which meant it had to come to Gorman to be ground first.

We ran the round bale proposal by Campbell management at a meeting in the Spring of 1970. At that time, I had the great fortune to meet Mr. Gary Vermeer from Pella, Iowa. He was the founder of Vermeer Manufacturing Company. He had the idea of rolling up hay in round packages of six feet wide and six feet in diameter bundles and had not been tried east of the Mississippi as far as we knew. Mr. Vermeer wanted to see this succeed for himself. He flew in on a small plane and landed near the mushroom farm in St. Charles, Illinois so he could promote the system one on one with Campbell management.

There was a point in the meeting where Mr. Vermeer was stressing how his hay would be preserved in these tightly wrapped bales and would not spoil for many months. Remember, Campbell's wanted the hay to deteriorate so it would be easier to grind and compost. The air in the room got very close and Mr. Vermeer quietly got up and started to walk out of the meeting. Gorman saved the day by making a suggestion that he could lay the bales on their sides after we baled them so water and the elements could infiltrte the hay and start the rotting process. Mr. Vermeer agreed that then and only then would his bales do what Campbell's wanted them to do and everybody left happy. I took Mr. Vermeer back to his plane and at the age of eighteen did not realize or dream what his company would become. He and Gorman were certainly innovators of equal caliber in my book. The company went on to be great inventors of hay equipment, a lot of farm equipment, grinding equipment and many other things that all received patents. I hear that the town of Pella, Iowa is almost all Vermeer from one end to the other. Gary Vermeer can certainly be added to my list of most respected.

Everything went well for about three years when we received a call on a very warm July evening that the piles of hay and stable bedding had caught on fire at the mushroom farm. After a three day fight to extinguish the fire, the town of West Chicago asked Campbell's to stop composting the way they had for the last twenty years for the fear of more fires in the future. At about the same time, a nearby drive-in theater started to complain about odors from the compost site. Probably one of the first air pollution nuisance problems from a compost facility in the U.S. In order to keep the farm going and producing mushrooms, the company started to research the use of enclosed rotary drums that were being tried on an experimental basis in a few parts of the world. Campbell's decided to install one of these monstrosities after very careful consideration and expense. The goal of this very expensive piece of equipment was to fill the drum on one end, and thirty hours later, have finished compost come out the other end, thus eliminating the two week outside process completely.

The drum was about twelve feet in diameter. It was over one hundred and fifty feet long. It sat on huge rollers that let it rotate about one revolution every two minutes. It was set on a five degree angle which meant it had to be loaded on the high end, about twenty feet in the air, with all the raw ingredients. There was an elevator that ran up to the top for loading purposes and a large building enclosed the entire process. The building was large enough to house all the raw ingredients that would be required for the next twenty four hours. The drum loading and turning continued around the clock, twenty four hours a day nonstop in order to feed the insatiable need for compost. The material would enter the top, and steam would be introduced so the material reached 170 degrees very quickly. The raw ingredients cooked that way for thirty hours constantly rolling down the drum a few feet with each rotation before it would fall out the other end and was called mushroom compost.

We adjusted our delivery schedule so there was no hay piles on the site anymore at the end of the day. The ground hay was now being sent at an "On Time" basis that is very popular today with a lot of corporations. Whatever they needed for the day is what we deliverd so the town would not complain about any outside hay storage. There has been a lot of research on what causes fires. 99% of the time it comes down to a cigarette or sparks from working on a machine and not spontaneous combustion as most fires are reported in the industry. Somebody was too lazy to put a cigarette out properly or someone left a machine with a hot spot on it to go home. All this reorganization at Prince Crossing because of a fire from an unknown source and thousands of dollars later they continued to grow mushrooms in West Chicago.

This type of compost operation was somewhat successful for about two years. An employee got crushed between one of the rollers that held the drum and was killed. Because of the insatiable need for compost, the man was pulled away from the site, and the drum was restarted within an hour. The local drive-in theater was still complaining to the town about odors and had filed suit. It seems that there was more odor from the drum than from the rick type compost made outside. Additionally, production was just not as good as it was with outdoor produced compost. Because of all these problems, OSHA and the town of West Chicago asked Campbell's to stop composting within the city limits. In order to continue to produce mushrooms at Prince Crossing, Campbell's decided to move the composting operation to one of their other farms in Glenn, Michigan. This newer facility built in the early seventies, produced rick style compost under roof for themselves and needed to about double production with the addition of the compost that was needed for West Chicago. Because of its remote location, there had not been any odor complaints from this farm. The problem was to pay for the 155 mile one way trip up to Glenn everyday with all the raw ingredients that were needed and the return trip of finished compost back to Prince Crossing. The decision was made and the old farm never skipped a beat, but the cost was high.

You must realize that the 155 mile trip from Chicago up along Lake Michigan to the Glenn farm goes through one of the worst lake effect snow belts in the country. This created constant headaches from late November to early March. The farm somehow always got their compost thanks to a very dedicated group of truck drivers, much tougher than any we have today. I knew of several drivers that used two different drivers log books and took two loads of stable bedding up to Glenn and two loads of compost back to Prince Crosing each day for months on end. Over six hundred and fifty miles per day without any rest but for big paychecks.

All of this was taking place during my high school and college days. I first entered a local junior college, Joliet Junior, the oldest one in the nation. I got rid of what I call those required courses before I transferred to the University of Illinois for my last two years. My goal was to take only agriculture related classes when I got to the U of I. While I was in junior college, I scheduled all my classes in the morning so I could work in the afternoon for Gorman. I took a combination of required classes along with a few agriculture classes the two years I was at the junior college and transferred sixty eight hours in the Fall of 1971 of a needed one hundred and twenty six to earn a BS degree.

In 1971, the U of I had over 1100 applications for transfer and become juniors from around the U.S. There was talk that only about 260 were going to be accepted that year. Although he has never admitted it, that same old agriculture teacher from high school intervened and I was accepted at the U of I in the Fall of 1971. He had graduated from there about fifteen years earlier and must have influenced somebody to take a chance with me. Again, no way to ever thank him for that. Not only did he get me that full ride at the junior college, I got a full ride at the U of I as well. That same old scholarship provided me with all the hours I wanted to take while I was there.

My studies drifted into the Agronomy area, mainly soils, but also into Plant Physiology and Agriculture Economics. What a tremendous group of educators I had while attending the university. There are still some I can call for conversation and see some at different college functions. All my classes seemed easy to me except one and one in particular. It was my second semester in the Spring of 1972 that for some unknown reason I took a Genetics class. This was the basic course that was semi-required in order to receive a degree in Agronomy. The class of eighty or so was made up of pre-veterinarian students with white shirts and ties, breif cases, and pocket protectors to boot. This class ended my fourteen year career of never getting any grade but an "A" and ended up with a 69.5 that was rounded up to a "D". The absolute toughest class I ever took and it really made me feel low when one of friends in pre-vet said it was his easiest class he had that semester.

I think I had the most fun attending the different games the U of I had in those days. The football team will go down as the worst in the hundred year tradition of the Illinois program. We still went to the games and had lots of fun and still attend a few after I graduated. The basketball was so-so, but the students got the best seats in the house and we never missed a home game in the two years I was there. Have gone back a couple of times since graduating, but tickets are harder to find and are much more expensive since they run a big time program now. I think we had the most fun at the hockey games. I was never much of a hockey enthusiast, but the venue was right next to our residence and the tickets were very cheap. The drinks were also cheap, and we did not have to drive back home. Enough said about that.

I lived with three other friends in this twenty story high rise building that housed about eighty people per floor. Our particular floor was comprised of everyone over the age of twentyone, so there was no limitations as far as alcohol and the floor was also co-ed. There was always some kind of party on the weekends on my floor and a lot of the younger people in the building were forever trying to get onto this level for some underage drinking. It was my job, being a little bigger than most, to make sure this did not happen. Somebody somehow slipped about six ounces of "Everclear" into a beer of mine on one particular weekend when my three roommates were gone. I got back to my room, opened up a window, and fortunately fell asleep for about twenty-six hours. When I finally woke up, the water was frozen in the room because it was about five below zero outside. It was a very scary situation now that I think back about how close I was to death from either falling out the window or from an over dose of alcohol. This happens a lot and many do not live to tell about it.

Shortly after I came home from my junior year in 1972, Dad passed away from a two year battle of colon cancer. I had made my mind up that I would not return for my senior year which would have left mom all alone. She talked me into returning and the diploma still hangs in the office proud to be the only kid from the family to graduate from college. You know, the rest of the kids were all successful without one for some reason.

Shortly after graduating, I met my future wife from the small town of Elwood. It was only about five south from where I grew up. She and I knew each other since we road the same bus to school everyday, but had not gone any further than that. Received a strange card in the mail which looked like an "S+H" green stamp that were so popular back in those days. It asked, "what was I saving myself for and if we could meet soon in Elwood?" Curiosity got the better of me. The next time it rained and we got off early from working in the fields, I drove down to Elwood and Debbie was at the the only gas station in town having something looked at on her car. It was also the only hangout in town as well. I pulled in and asked her about a certain card I received in the mail, and she denied ever knowing anything about it even to this day some thirty seven years later. We hit it off right away, and were married a year later. Whenever I bring it up, she still denies ever knowing anything and tells me when I turn sixty, she is going trade me in on two thirty year olds.

From the day we first started to date, she has always been behind me one hundred percent with all my endeavors. After she miscarried four times early in our marriage, we went to Chicago to investigate the possibility of in-vitro fertilization at a cost of about $25,000. The appointment happen to fall on my birthday. When we arrived back home, the phone was ringing. It was a call from her OB-GYN telling us that he had a baby that was going to be delivered in the month of June. He wanted to know if we were interested and if we could take over the bills for it to be born? We could not say "yes" fast enough.

We followed the young girls pregnancy for the next five months, even hiring a cab to take her to doctor appointments when she had no way of getting there. A baby girl was born in early June, and three days later we were able to see her for the first time. It was such a rare event to see an adoption in that particular hospital. There was an audience of at least twenty nurses and doctors sending us off with our new baby Kimberly. My mother-in-law was able to come up to the house and help out the first two weeks and made our transition to parenthood very easy.

Family Growers And Campbell Soup

In the Spring of 1973, three small family mushroom growers from the Chicago area approached Gorman to see if he would custom compost material for them. At one time, there were over twenty five mushroom growers in the greater Chicago area. At about the same time, our portion of ground hay going to Michigan had dwindled down to nothing because of the transportation costs involved. We no longer needed the WHO tub grinder. We cleaned the machine up and repainted any areas that needed it, and advertised it back out west. In a month, the original manufacturer,WHO, called and made an offer to purchase the machine if we could deliver it to Montana within the month. We used it for four years and sold it back for $58,000 or about one thousand more than Gorman had originally paid for the machine. I am still in awe of that machine even after forty years and what it could do with a three thousand pound round bale of hay.

If we made the decision to compost for these small growers, we would now have some money to buy the special equipment needed to do the job and lay down some more concrete. Since Prince Crossing was having their compost made at Glenn, Michigan, the machinery they once used was just sitting at their old compost site rusting away. Gorman had to make three important decisions if we were going to do the custom compost job. He would have to pour about two more acres of concrete to the one acre pad we had for grinding hay, we would have to buy some special equipment that we hoped to get off of Campbell's at a good price, and somebody had to run this thing seven days per week. He easily made the first two decisions without a problem and he turned to me for the final decision. I of course said yes after I saw the operation at Prince Crossing and became addicted to composting back in 1968. I graduated from the University of Illinois on June 2, 1973 at 10:30 AM with my mom, my older brother, and his wife in attendance. By 3:00 PM, I started to make my first batch of compost for three small family operations on Gorman's farm.

By this time, a lot of the land in the Arsenal was no longer restricted to growing only non-government supported crops. The farm operation balloned to over five thousand acres, we had three semi-trucks on the road hauling gravel, and grain, and we now added this compost operation to the mix.

Before we began making compost for these small growers, Gorman sent me out to Kennett Square, Pennsylvania, the mushroom capital of the world. I was going to nose around a little to see what I could pick up as far as the process of composting and how to price such a venture. The towns of Kennett Square, Avondale, Toughkenamon, and Kaolin are all located about forty miles northwest of Philadelphia, Pennsylvania. There was a large influx of Italian settlers into this area and many started to grow mushrooms like they had in their home country You will usually smell a compost operation before you see it, so there was no problem with directions. The area was filled with custom composters and mushroom houses everywhere you looked. I drove into the first site early on my first day and had the good fortune to meet Jim Frezzo and his family. A short, husky, soft spoken Italian who was in charge of a massive custom compost operation and about fifteen employees. We were contemplating producing about two hundred tons of compost per week. Jim and his family were producing over two thousand tons per day. Yes, that's right, two thousand tons per day!

After I introduced myself and where I was from, Jim had no apprehension about me since I would not be in competition for any of his customers anytime soon. Jim and his family took me in like I was one of their sons and spent almost the entire day with me pointing out all the do's and don'ts with making good mushroom compost. He let me know a very simple formula for pricing the finished product and have used it my whole life. He said, "add up all the cost of your ingredients, and then double that cost for the price of finished material." After he had spent most of the day with me, he invited me to go with him that evening to the annual frog leg fry where most of the growers and custom composters would be. I had to turn him down and still get sick just thinking about eating a frog leg forty years later. He and his family can certainly be added to my list of most respected. I have been back to that area about three other times since then and always stop and see Jim and pay my respects.

On the three day trip, there were far more operations that would not even let me set foot on their property and ask a single question. On the last day of the trip, I stopped at the Kaolin mushroom site. I met Mr. Lou Pia, who from my conversations with other composters, appeared to be the "Godfather" of the custom composters. My foot in the door was to ask about some of the machinery out front that was always for sale at an operation of that size. He was chasing a fly with a swatter in his office as I introduced myself, a problem anytime you have that much stable bedding or compost around. He wanted to know my name and asked, "if I wanted to see the machinery or wanted to learn how to make good compost?" To this day, I think Jim Frezzo had called Mr. Pia and said, "watch out for this greenhorn and do what ever you can for him." Lou introduced me to his son's, John and Mike, who showed me everything they could in a two hour compost lesson. I want to thank the Pia family and the Frezzo family for their hospitality and will forever be grateful.

Just before I left the Kaolin site, I stepped in what I thought was a thin film of compost water only to find out it was a hole about a foot deep. Since I did not have a change of shoes, I tried to wash the brown water off my pants, socks, and shoes the best I could before driving back to the Philadelphia airport. I checked in and got on the plane as far back as I could and to my fortune a young girl with a baby sat in the same row. Through out the two hour flight, the girl kept checking the baby figuring it had done something in it's diaper when it was actually my stinking feet the whole time.

I took all the experiences from the trip back to Gorman's and after graduating, we started to develop this custom compost operation. That old choke feed turner I had seen many years earlier was purchased for five thousand dollars from Prince Crossing. The engine was a total disaster, so Gorman purchased the cutest little John Deere engine that fit right in the place of the old engine and worked like a charm for many years. The machine itself was very rusty, but after a few days of material going through the hopper, everything shined up very nicely.

We started to put compost on the market in the summer of 1973. We used the same formula as Campbell's did back the first time we visited the farm in 1968. We once again had an outlet for old or damaged hay, we added two positions to the work force, and we definitely had year round work. The basics again for mushroom farming, is a special material called compost is prepared and filled into special growing houses. Mushroom growth and harvesting occurs over the next ninety days, then the old compost is removed and is now called "spent mushroom compost", and a new batch is prepared and filled again. Those ninety days of production and what needs to done could fill a small library with all the research and publications that have been compiled in the last one hundred years.

One thing that must be accepted when your customer is a mushroom grower and you are the compost provider, you will never receive a compliment. It is standard operating procedure in the industry. You will only receive comments that usually include the material was too wet or the material was too dry, it was too green or too overdone, they had too much material or they did not have enough material. If the grower got a good crop of mushrooms, it was what he had done, and if he gets a bad crop, it was what I had done. I know of custom compost operations that have over two hundred customers for the same material and the production will vary by one hundred percent from one grower to the next from the same product. We used to laugh about all this because we knew if we did not hear anything, it must have been very good compost.

Shipping two hundred tons off the farm each week, year round, meant we had to make compost outside during those miserable Chicago Winters. There is not enough pages in this book to describe how bad it can get in January and February in northern Illinois. The amounts of snow were bad enough, but the greatest problem was the wind chill. I don't know many people that have worked in sixty-below wind chill and actually accomplished something. We did survive, we did produce quality compost with that old choke feed turner and a loader without much heat, and we looked forward to warmer days.

During those first few years of composting, Gorman was able to expand his grain operation to over five thousand acres. Much of the Arsenal land went up for bid every five years, and all the land had now dropped the stipulation about not growing government supported crops. It attracted a lot more interest and cash rents started to go up drastically. The land we rented in 1975 for forty five dollars per acre, brought more than $110 per acre in 1980. Things were about to change in a large way for all of us.

In June of 1978, Campbell Soup management approached Gorman on a mission. They wanted to cease composting the material at Glenn, Michigan for the Prince Crossing farm in West Chicago. The proposal was to have us start making the compost on Gorman's site and deliver it to Prince Crossing for mushroom production. Fuel prices had sky rocketed and there was a lot of labor disputes with the truck drivers. They were again faced with a decision to try and keep this farm in West Chicago growing mushrooms at any price. They were aware we had been making compost for some of the small family operations around Chicago and were willing to give us a try. Gorman confronted me with the chance to take this two hundred ton per week operation to one thousand tons per week. It was a very easy decision for me because of my addiction to compost. Gorman made an offer using that old formula I got from Frezzo some five years earlier with the stipulation that Campbell's would supply the needed turning equipment for a job that size and we would maintain it.

The offer was given to Campbell headquarters in Blandon, Pennsylvania and of course, it was rejected. Jim Elliot, the man in charge of contracts stated that, "there other five farms were producing compost for much less than our offer and to resubmit a better number." Gorman held to his guns on my advise, a twenty seven year old telling a fifty year old what to do, and about a week later our original offer was accepted by Campbell management. We held to our original number from the advise I got from Jim Frezzo who reminded me that we had expeience in making compost, we had a site already to go, and our location was excellent. Raw materials started to arrive, some used self propelled turning equipment was delivered, and we began supplying Prince Crossing with compost on July 4, 1978. A five fold increase off the farm, and the introduction of all that newer equipment sure made the job a lot easier.

That first full year was of course the hardest. A lot more man hours, a lot more men, and a lot more headaches, but my biggest concern was the first winter with all this extra material. Campbell's did not let us have all the production for Prince Crossing until May of 1979. They wanted half of the material needed to come from us in Joliet, and the other half from the farm in Glenn, Michigan. They could then compare which compost got the best results. I have always considered outdoor produced compost to be superior to indoor produced compost. Despite everything mother nature threw at us over the ten month experiment, the compost we made did produce a few more mushrooms than the compost from Glenn at about half the cost.

The weather that first winter with Campbell's was of course for the record books. The tempeature dropped below freezing on December 8, 1978, and did not get above freezing again until February 19, 1979. Many days it was well over twenty below zero with wind chills at fifty below. If the temperature was not bad enough, the amount of snow was paralyzing. We made the compost ricks about eight feet wide, about seven feet high, and about one hundred and seventy five feet long all outside on this two acre concrete pad. Because of our limited space, we built the ricks about three feet apart. This was just enough room to allow the self propelled turners to run through the ricks and provide the needed oxygen and water for proper composting. On many occasions that particular winter, the snow came in so heavy and blew so hard that you could not tell what was a rick of compost and what was snow. All of those three foot alleys between the ricks filled in level, over seven feet deep, with snow. We had to shovel all the snow out by hand before we could run machinery down the ricks for turning. We ended up far more times than I want to remember cleaning these alleys by hand bringing the snow to the ends so equipment could move it off the concrete. One of the biggest reasons I have back problems today, some thirty years later.

Another lesson we learned during that winter was that compost at a 170 degrees dumped into an aluminum trailer and hauled forty miles will not slide out when it is less than fifteen degrees outside. Our first taste of this came on a day when we tried to deliver material to West Chicago when it was fourteen below zero. The truck driver raised the box and not a single pound of compost came out . We all went up to the farm with forks and shovels to unload the trailer by hand. Gorman decided that with the next load we would first put down a piece of plastic on the floor before loading. The next load came out like the compost was sitting on Teflon. The people at the farm had very little trouble removing the plastic from the pile and were very happy to receive the compost and fill their house. The total cost was about a dollar per load and we used this trick every time the temperature dropped below fifteen degrees.

In 1980, Campbell's made the decision to stop growing mushrooms for the soup company and start growing mushrooms for the fresh market. Up until that time, mushrooms were considered a delicacy and not many dollars were being spent for them. That image was changing and consumption of fresh mushrooms across the U.S. increased sharply. They increased their quality control on the product, picked and packed the mushrooms in one pound containers, and delivered to the local super markets all in less than twenty four hours. This decision would eventually lead to the demise of all the small growers in the greater Chicago area. Instead of the housewife going out of her way to a small mushroom farm for her needs, she could now pick up fresh mushrooms at the grocery store along with all her other needs. Campbell Soup was so efficient that if a package had not sold within four days, they would replace the old package with a new package of mushrooms and sell the old product to the soup plant. The retailer was out nothing for the replacement. By 1990, there were only four family mushroom growers left within 165 miles of Chicago from the original twenty-five.

In addition to that, Ralston Purina Company decided to enter the fresh mushroom market by building a facility twice the size of Prince Crossing near the town of Princeton, Illinois in 1976. They provided heavy competition for fresh mushrooms in the Midwest, and competition for all the raw ingredients needed for compost production. In a usual week, there was over a million pounds of mushrooms that had to be absorbed in the upper Midwest in the late seventies.

Ralston Purina was certainly know for animal food but not much for human food. They found and purchased about 160 acres of land near Princeton, Illinois. They also built five other farms through out the U.S. in a period of less than three years. The farm in Princeton was located about ten miles off of Interstate 80 in the middle of no where. The main ingredient for most mushroom compost was stable bedding and this farm was located over a hundred miles from the nearest racetrack. Competition for stable bedding became outright crazy with Ralston gearing up with trucks and trailers to haul the bedding away from the tracks, and at one point even paying the tracks to get it. Up to that time, the vendors removing the stable bedding were paid handsomely for the removal since it was a waste product and there was only the small growers and Campbell Soup that needed the stuff. To make matters even worse, Washington Park, a large racetrack near Gary, Indiana burned down in 1979 and was never rebuilt. During all this turmoil, we were constantly trying to acquire enough material to keep our small growers going. We scratched by using more hay and more cobs in the formula and came up with some very productive compost that Campbell's were willing to try and save some money.

In 1978, the trucking by Ralston Purina went belly up, and a man named Fred Noorlag came in and bought all the equipment for about twenty cents on the dollar. Ralston signed contracts with Fred to have him deliver a certain amount of material to their farm each week and not a pound more. The horse racing industry was at its peak and stable bedding was coming out of the seams from all five tracks. Ralston was paying for it when they should have been getting large sums for removing the material. All the track contracts were renegotiated and Fred started to receive money for the cleanup, loading, and removal of the stable bedding within three months after he purchased all the machinery from Ralston Purina. Fred brought in his two son-in-laws to help with the operation, and we started to receive stable bedding at Gorman's from Fred in the fall of 1978. Little did I know what a lasting relationship this would become with Fred.

The material was being produced in such large quantities in the summer of 1980 that Fred asked Gorman if he could stockpile excess material on his farm in Joliet. He brought in more than five hundred loads of stable bedding that had no home. When it was dumped, we pushed and piled the material as high as we could with the machinery we had. Ralston had adjusted their formula to include thirty five percent wheat straw and only sixty five percent stable bedding which eliminated the need for about three thousand yards of race track material per week at the same time there was abundance in Chicago. Just before Thanksgiving of that year, the old Glenn, Michigan farm called Fred and said they were desperate for stable bedding and could use two hundred loads if the price was right. Fred just so happened to have some excess at Gorman's and the price was right.

Fred would have his trucks complete their job for the day in Chicago and then show up at Gorman's at about 1 PM each day to be loaded for the 155 mile trip to Glenn with a load of the old material. Gorman had purchased one of the best pieces of loading equipment anywhere to load dry loose material. It was called a "Haymaster", out of Garden City, Kansas. You could pick up ten yards of loose material with its large fork and grapple and easily load over a thirteen foot trailer without a problem. It was equipped with four wheel drive and sat on four large flotation tires so mud or soft ground was never an issue. The trucks would pull in and twelve or so big bites of stable bedding where loaded in, the load would be tarped, and off they would go. The Glenn mushroom farm liked the material so well, they took every pound of excess material that Fred had dumped on Gorman's farm that summer.

Fred hired two young brokers to help haul in addition to his own trucks. These two kids worked twenty four hours a day for over a month. If they needed to get loaded at three in the morning or at six at night, someone was there for them. I understand that these two kids went on to build their own business called "Mr. Bults" or MBI out of Chicago and have over a thousand trucks hauling garbage in six states. Fred paid Gorman very handsomely for all his help with the excess material, and Gorman paid all of his employees a nice bonus that Christmas of 1980. That's the way Fred and Gorman were.

By May of 1982, everything was going very well for the Gorman O'Reilly operation. The Arsenal changed farming policy and you could grow whatever you wanted. Gorman had acquired over five thousand acres of land. I had also bid on and got over one thousand acres for myself. I used Gorman's manpower and machinery to farm my ground since I first farmed some of my own land in 1974. It was an arrangement that worked extremely well. We had three semi's on the road carrying compost, gravel, and grain. We had this very large custom compost operation for Campbell Soup and the small growers around the Chicago area. Things were about to change in a very big way.

Death, Farming, And The Ladies

It was May of 1982 and everything was going great. It was an early Spring in Illinois and we had all three thousand acres of corn planted by the ninth of May. We always took very good care of all the equipment, so on the tenth of May, we had everything that Gorman owned back at the home base to change oils and grease before we started to plant soybeans the next day. It was now the eleventh of May and Gorman left the farm to get some half and half which seemed to be a cure-all for his physical problems. I thought it had to be an ulcer from all the stress of managing that many men, that many acres, and that much money. I was in the little farm office we built about a year earlier to house machinery records and the two way radio equipment. He called on the "FM" radio to let me know we needed six trucks to go to West Chicago that day instead of the usual five. When I acknowledged him, he failed to say ten-four back which he never failed to do. I tried to respond to his call again, and still no ten-four reply. I looked at the clock and it was 8:15 AM. About a half hour went by, and a Will County police officer drove into the yard and informed his wife that Gorman had died in his car about a mile from the farm ending up in the road ditch.

It was an apparent heart attack at the age of fifty four. Tragic for his wife and daughter, tragic for his family, tragic for the farm community, and tragic for me. I will never forget the day we buried him. It was planting time in Illinois and most of the farm community, even his very close friends, thought it was more important to plant than attend his funeral. I can't remember how many times Gorman would leave the field early, go home and get cleaned up, and attend such functions in the seventeen years I worked for him. If any of his employees had the need to do the same, there was never a problem.

It was such a shock because we all thought he was in tremendous shape with all the hard work performed all those years. His own dad was almost ninety when we lost Gorman. He died leaving his wife and daughter in charge of this massive operation. Two individuals that never showed any interest in the farming or composting operation in the seventeen years I worked for him. They, in my opinion, were nothing but argument and may have been the reason he worked so many long hours away from home.

Gorman was always buying and selling John Deere equipment trying to make his line of farm tools better and more efficient. I was told by a very reputable equipment dealer he worked with for over twenty five years, that in the Spring of 1982, Gorman traded for but had not taken possession of two new combines, two new corn heads for those combines, two new soybean platforms for those two combines, a new rotary hoe, and a four wheel drive tractor. The purchase and trade price was in excess of seven hundred and fifty thousand dollars. That combined with other outstanding loans from Deere totaled almost two million. This was wiped off the books because Gorman had John Deere Credit Life Insurance on all his loan dealings which cost next to nothing. All the equipment on the farm and all that new equipment to be delivered was free and clear.

In January of 1982, Gorman applied for a million dollar life insurance policy with Country Companies in Will County, Illinois. The insurance company made Gorman give blood and take a stress test before insurability was established. He of course passed with flying colors. He made one small payment in February of that year and three months later he was dead. In the first couple days of April 1982, Gorman wanted me to see a check paid to a local bank that represented the last payment on the farm he purchased some twenty years earlier.

So his wife and daughter began their farming career with all the machinery free and clear according to that local John Deere dealer, the farm where the home base of all operations was established was free and clear, had over one hundred thousand bushels of grain in storage from the 1981 crop year that was not sold, had all the corn planted for 1982, and had a million dollars in the bank. I know that many of Gorman's friends tried to convince the two ladies to sell that Fall and not attempt to try and run such a huge operation. They did not take any advice, and the big change in management was under way.

We picked ourselves up in a couple of days and got the rest of the crops planted, continued trucking what needed to be moved, and prepared and delivered the compost for Prince Crossing. This is were I should have had things in writing. With a hand shake between Gorman and I, he told me that on the fourth anniversary of making compost for Campbell's, July 4, 1982, half of the net proceeds of the compost operation were going to be turned over to me. He said everything was paid for, the extra equipment, the extra concrete, the scale, the trucks, and he thought I deserved the bump in pay for keeping the small growers and Campbell's happy on a seven day per week basis. When I discussed this with the two ladies late in June of 1982, they had different ideas about my income. This was the first time I told them how unhappy I was working for them, and I let them know that February 1, 1983 would probably be my last day if things did not change. By then, the 1982 crops would be harvested, and most of the Winter would be over with concerning compost production during the cold months of the year. I got zero response from my threat of leaving. It was like everybody can be replaced and they just brushed it off.

One of the first things that happened after Gorman died was to hire a non-farm background CPA to help them manage the operation and money. The week after he was hired, he approached me and said Gorman was paying me way too much for my seven day per week services. Everybody got their wages reduced the first month by about twenty five percent. He and the two ladies also made some horrendous decisions on selling grain stored on the farm from 1981 crop year, and selling grain that had not been harvested yet in 1982. I don't believe Gorman ever speculated on any grain sales to my knowledge, if he had, it would have been a more educated decision that what you are about to read about.

A good example of speculation occurred late in 1982 and early1983. Corn approached the $2.60 per bushel range, which was a lot better than it had been for a very long time. From what I was told by the local elevator, the trio sold a hundred bushels of corn per acre that was not even planted yet in 1983. Soon after they made the sale, the government announced the "PIK" program which paid farmers not to grow corn in 1983. A lot of bad publicity about farmers getting paid to do nothing is in the news everyday. This was one of those situations where it actually came true. A lot of farmers agreed not to plant corn and the price rallied to over $3.85 per bushel in our area. The trio thought it was more advantageous to plant and raise corn and to spend all that money on inputs than to take the government program and raise no corn and have no inputs. 1983 was also a drought year in our area, so any corn that was produced was very poor as far as yields. Most of the corn the ladies harvested came in at seventy five bushels per acre or less, about twenty five bushels short of the contract. They went out and had to purchase the shortfall at $3.85 per bushel or about a half million dollars. The corn that was produced was sold at $2.60 or a shortfall of another half million. So the one million life policy Gorman had was quickly ate up after just one year of farming.

They made a decision to have a huge farm sale of all Gorman's equipment on December 7, 1984, a day that will live in infamy. The two years they tried to farm, 1983 and 1984, the machinery had very little upkeep or maintenance because the CPA figured it would be cheaper. The day of the sale, most of the machinery was considered in rough condition and brought about three quarters of a million less than if it would have been sold the fall after Gorman died.

I will describe one of the insidious things that occurred the day of the sale. There was one tractor that would not start. One of the employees sprayed so much ether into the engine that when it tried to start, the engine block cracked right on site. Needless to say, that tractor did not bring much money. I can only imagine how it would have effected Gorman to see all his beautiful equipment sell and his beautiful operation go down the drain. It sure made me sick. It also made me sick to see all his friends there. Very few could come to his funeral two years earlier because they thought it was more important to plant, but they all had time to see his machinery sell. I also got the impression by many that knew Gorman that it was partially my fault everything went to hell because I let the operation some two years earlier. I can't say if I would have made any difference, but it was the two ladies who made the decision for me to leave with my mind free and clear.

After I left the ladies and the operation in February of 1983, I was not completely done with them. I had acquired about 1100 acres of land over my seventeen years with Gorman and used his manpower and machinery to farm this land. It was kind of a trade off for me working seven days per week with the compost business. Most of the time I never got to see my land as I was too busy composting. The trade off worked well while he was alive. We unfortunately did everything with a handshake. I would normally meet with Gorman for five minutes in December to settle up for that years expenses from using his equipment and manpower. His answer for the previous twelve years was to give him a check for $10,000 and we will call it even. It should have been more, but he realized what was being made off the compost and a check for $10,000 was more than sufficient. Since my acreage had not changed from the 1981 season, and we were still putting out as much compost as ever, in December of 1982 I presented a check to the two ladies for $10,000. Santa arrived early that year and they presented me with a bill for $39,000 for use of machinery and manpower to farm my land.

There are hundreds of publications and guidelines from Midwest universities that give farmers some idea on what it cost to do anything related to a farm operation. So much to plow or disk an acre of land, so much to harvest an acre of soybeans, and the list goes on forever. Gorman and I both knew of these charts but he never used them in the past. He figured my worth to the farm, deducted that from the bill, and $10,000 seemed to be a very good number for all of those years. Needless to say, I did not pay there bill and another reason why I left the operation in 1983.

Gorman always thought it was best to plant or harvest like an army. Start at one end and work to the other end. It was the most efficient and made the most sense. In the fall of 1982, we were harvesting some of the ladies land which happened to be right next to my property. It would have made common sense to jump into my land while we were there and harvest the crop. A decision was made to jump more than fifteen miles and continue harvesting their crops even though it required a day to move all the equipment. That would have never happened with Gorman and even the employees knew things were going to be different from here on out.

On January 12, 1984, I was relaxing at home in anticipation of going out that evening with my wife for my birthday. A knock at the front door and upon opening it revealed a Will County police officer. He served me with a summons to appear in court and pay the O'Reilly ladies the remainder of my bill, $29,000 plus interest for the use of Gorman's equipment and manpower in 1982. Seventeen years of service to that farm and Gorman, most of it seven days per week, and it meant nothing to these two women. I could not believe it. After coming to my senses, I called and got a farm background lawyer from Wilmington, Illinois named Bill Francis. They got an expensive lawyer out of Joliet and we all filed into court in July of 1984 for a two day bench trial. Many witnesses were called by their side all eluding to the fact of these university custom rate numbers and how expensive it was too farm in those days. My lawyer chewed up everyone one of their witnesses with just a few questions mainly referring to the fact that not a one of them ever farmed in their life or paid a bill relating to farming.

My lawyer called just two witnesses, a farmer that helped me out with some of the harvest in the Fall of 1982. I felt the ladies were going to let my crops go to the end and harvest could be hampered by the weather. And he called on me to testify. The farmer, a witness that actually farmed and did some custom harvesting for me, stated much different numbers than what the two ladies were using. The two ladies also included a lot of crop acres in their bill that the farmer I hired actually harvested and trucked to the elevator which made them look rather ridiculous. When I was called, my lawyer only asked me about what I had paid Gorman over the last twelve years? Because of my FFA days and meticulous record keeping, I was able to answer and produce every check I had paid Gorman for the last twelve years. I was able to give exact acres planted and harvested and trucked on my 1100 acres for all the prior years.

After we had finished testimony the first day, their lawyer approached our lawyer and offered to settle for $15,000 or just less than half of what they were asking for. We had not even put our case on yet, but because of the way my lawyer was eating up all their witnesses, they were willing to take less. Bill did tender the offer to my wife and I, and with his advice, we declined the offer and pushed on with the trial the next day. After the two day trial and closing arguments, the judge said he would have a decision by the end of business the next day.

We could not sleep that night and were on pins and needles the entire next day. Around 4:30, our lawyer called and said, "no money was to be paid." We were speechless. The judge did give us five very good reasons for his decision. The first, nothing was said to me after Gorman died in May that anything would be different than it had been the prior years. The second, none of their witnesses had ever been around a farm and really knew what was charged for machinery and manpower. The third, the farmer that testified for me had charges much different than the experts had told in court and the acreage was very different than they were charging. The fourth, because of my records, it looked like the going rate for that amount of acreage should be $10,000. The last reason was I wrote on the check, "Paid In Full". Even though they did not cash the check until ten days before the trial on the advice of their lawyer, the judge consider it payment in full for machinery and manpower for the year of 1982.

They had thirty days to appeal the decision of no money was to change hands. They did appeal on the last day close to 3:30 in the afternoon. The case was sent back to the same judge that sat through the first trial. My wife and I, my father-in-law, our lawyer, and their lawyer were the only people to show up for the appeal that day. After about an hour tirade from their lawyer over all the aspects of what he thought was an improper ruling, our lawyer got up and expressed the fact he thought the ruling was fair and pointed out that my wife and I were the only ones to show up for the appeal that day. The judge quickly decided in our favor again and in addition to that, placed sanctions on the two ladies for taking up the courts time and our time. He awarded the original $10,000 I paid back in 1982 to be paid back to me. They never paid the money and I never went after it. My Dad always said, " pay your fair share," and I felt I had paid my fair share. My lawyer fees were about $400 dollars and have used Bill Francis on many occasions since that day. He can certainly be added to my list of most respected.

In the Spring of 1983, I started to farm for myself. I bought some farm equipment, all John Deere of course. A couple of used tractors, an old disk, a small field cultivator, and a corn planter. It almost felt as good as the day I ran through all those lineman back in high school. It was extremely hard to adjust without Gorman being around. I still had the composting addiction that was cut off cold turkey when I left the O'Reilly operation in February of 1983. Things were about to change again in a very big way.

Compost Products; Birth And Death

It turned out to be a dry Spring in Illinois in 1983. I had accepted the PIK plan from the government not to plant any corn for that year. Instead, I took the government check and did not have to pay all those crop expenses or worry about the weather that year. I did plant all the corn acres to a cover crop called "Sudax", which grows to about twelve feet tall resembling savanna grass in the jungles. It provides a lot of organic matter to the soil and requires very little moisture to grow. It was an excellent choice for that very dry year. Because I still had that composting addiction, I made a call to some old friends at the mushroom farm in West Chicago. Bill Edholm was the farm manager and talked to him about ten minutes to see how things were going after being gone for some three months. He indicated that things were not going very well. Communications with the two ladies were very poor, the quality of the compost had significantly dropped, and that Campbell Soup was considering not extending their contract which was up in September. He suggested that I should give Fred Noorlag a call.

I had known Fred for about five years. He supplied the stable bedding for the operation at Gorman's for many years. He had all the contracts for all the tracks in Chicago tied up for a long time after Ralston Purina stopped hauling the material. His background was in the garbage business and he treated the stable bedding as garbage. There was a cost for the removal and disposal. His great talent was to acquire small garbage operations, clean them up, expand their business, and then sell them to Waste Management for shares of stock. He was from a Dutch background, and his word and handshake was good as gold. It was my good fortune to meet both He and Gorman so early in my life. They were both self made men and had impeccable reputations. The day Gorman died, Fred was the first person to come to the farm and ask if there was anything he could do. After talking with Edholm, I immediately called Fred.

I took a few minutes to catch up with Fred and before long he asked if I wanted to run a compost operation for him. There was never a problem hearing Fred. He thought everybody had a hearing aide. He let me know that Prince Crossing was not happy with the whole situation in Joliet and was asked if he would be interested in setting up a compost site to do their mushroom compost preparation by the September deadline? He let them know he was interested, but the timetable seemed to be impossible to do. Fred invited me up for a one on one talk to see if it was possible. It was planting time in Illinois as I was trying to plant my soybeans for the upcoming growing season. I felt a little like Gorman when he would leave the field, get cleaned up for some wake or funeral. I decided I would go home, get cleaned up, and go see Fred about my addiction, I mean the new compost venture.

I found out Fred had already purchased eighty acres out in LaSalle County on a main state road about twenty miles from my home. It was going to be called Compost Product's, Inc. He asked what I would need to run this operation for him? I shot out the figure of $500 per week and he immediately shock my hand and I was hired. Little did I know it would be my home for the next twenty seven years. I did ask Fred to have all my duties, compensation and benefits spelled out in a contract. He had absolutely no problem with that and something I sure wished I had some twelve months earlier with Gorman.

The most important part of the contract was to keep Campbell Soup happy, they were the boss. Other than a few small growers we agreed to supply, Prince Crossing was are only customer and we had to keep them happy. In a corporation the size of Campbell Soup, praise was never a part of their vocabulary. The better the farm did, the more they wanted and the more was expected of you. We ground hay for more than four years early in my career and never really satisfied Campbell management one time, or if we did, we never heard about it. It was a two day job removing and replacing those hammers in the tub grinder trying to please Campbell's, but we never did in four year's.

So, on May 15, 1983, I signed a contract with Fred Noorlag to run Compost Product's and at the same time came to an agreement for time to farm my 1100 acres of farmland. As it would turn out, I stopped farming in the Fall of 1986 after three straight droughts in my area and four very tough years after Gorman had died. I was working past eight almost every night. I think they call that burning the candle on both ends, and I was sure feeling the heat. I let all my landowners know of my decision to stop farming, and I think it took less than twenty four hours for all the land to be rented out to new farmers. I sold off all my machinery to a single individual, had a few dollars left over, and paid the IRS over $42,000 for recapture taxes on the farm equipment. I had to pay the government to stop farming. You don't read about that in the paper everyday!

We broke ground for Compost Product's on June first with Fred depending on me for every decision that had to be made. To my amazement, every time I told him what was needed, he would normally double the request. We could use three acres of concrete, he poured five acres. We could use a small office of five hundred square feet, he built a thousand square foot office with all the amenties you would want. We could use a fifty ton scale, a hundred ton was installed. I wanted about fifteen thousand square feet of heated shop for the turning equipment, he built over thirty thousand. I wanted two good John Deere loaders, he bought four.

This went on for two years and I think well over one million was spent from the time we broke ground.

The facility was a showcase and model for any future compost site in the U.S. The EPA would use these specifications whenever a project was presented across the U.S. Drainage of excess water is the main concern of any compost operation. Our concrete pad drained very well because of a one degree slope north and south from the center. It also drained from east to west with a one degree slope. All the excess water would drain into a zero discharge retention pond that could hold a hundred year rain event. In my twenty seven years at the farm, we had over six, one hundred year rain events.

Mushroom compost requires a tremendous amount of water to properly compost all the dry material you initially receive for mixing. We were able to set up an irrigation system that pulled water out of the retention pond and water the incoming materials. It cut our fresh well water usage about 99% in the total system. Before we implemented this system, it was nothing to use over fifty thousand gallons of fresh water per day for the process. We also built a large covered warehouse open to the east that was divided with concrete walls to form ten individual bays to keep all the different supplements dry and separated. We bought all the nitrogen supplements like cottonseed meal, chicken manure, and urea in bulk and other ingredients like gypsum and sugar beet pellets by the semi-load and stored them under roof to keep them dry until use. One of the worst smells on earth is cottonseed meal that has gotten wet and sits out in the sun for a couple of days. Nothing else even comes close, believe me.

The concrete pad was laid out on top of very hard clay and compacted. A six inch layer of one inch stone was laid down and compacted. The concrete was poured at a six inch depth with heavy wire mesh. I tell you all of this because of the one hundred compost sites I have been on, most suffer from potholes and cracked concrete from the very heavy machinery that does most of the turning and handling in compost production. You may remember the story of me stepping in the one foot hole on the Kaolin site back in 1973. Our concrete pad, even after twenty seven years of heavy use, shows nothing and looks like it was just poured about a year ago because of all the site prep we went through before installing the concrete.

The big hurdles of producing compost by September were quickly being over come. We had a concrete pad, we had the management, we had some manpower, but we did not have any turning equipment. That sat forty-five miles away on the O'Reilly farm and they were still using most of it to finish off the contract. The actual turning machinery was always owned by Campbell Soup from day one. That's the way the contract was set up. We were responsible to maintain the equipment to insure proper composting would be completed daily. Since I was away from the farm for now six months, I did not realize how beat up and tore up the machinery had gotten in that short time. Compost is hard enough on the machinery and not keeping it maintained just adds to the misery.

A date was set to move half of the turning equipment to our farm in LaSalle County so we could start to make compost and the filling schedule not be interrupted. It would require a large crane, a couple of lowboy-semi's, and some expertise on how to handle the actual loading procedure. I returned to the O'Reilly farm with the moving equipment. About the time we were ready to load the first turner, the two ladies got wind that I was on the farm and were going to run me off with a shotgun if I did not leave immediately. I don't know if they had a gun or not, but I did leave without confronting them. I had to try and orchestrate the loading from the country road that ran by Gorman's property some eight hundred feet from the loading area. I guess you could say that not only were they mad they lost the contract with Campbell Soup, but they were even more furious to learn that I was the one in charge of the new operation. When we returned about two weeks later to retrieve the rest of the equipment, I sent one of Fred's son-in-laws to oversee the move.

We successfully got all the Campbell machinery moved the forty five miles and started to make mushroom compost. The two O'Reilly ladies asked Fred if he wanted to buy their two John Deere loaders that they owned for our new facility? After seeing the turning equipment and the shape it was in, I advised Fred to turn their offer down for two Deere pay loaders.

Campbell Soup had about ten thousand yards of stable bedding left on the O'Reilly farm in Joliet that Fred wanted to move to the new facility in LaSalle County. Before we got a chance to start moving this material, we had one of those winds out of the southwest that gusted over forty miles per hour for about two days and two nights straight. For some unknown reason, the stable bedding caught fire and the fire department could not extinguish. That sure would have made some good mushroom compost. Campbell Soup was officially done with the Joliet operation that showed so much promise the day we shipped the first load back on July 4, 1978.

What a hectic pace for those first four months. Putting in and permitting a new first class facility, teaching the manpower a little about composting, and with a million dollars later we delivered our first house of mushroom compost to West Chicago on September 1, 1983. Because of the amount of effort on my part to successfully accomplish this goal and deliver product when we said, Fred took my contract and crossed out the number $500 and inserted the number $700. What a great way to start my next twenty seven years and to feed my addition to compost for a very long time.

The first year I farmed by myself was also one for the record books. I did not have any corn that year because of the government PIK program. I did have about 500 acres of soybeans. On June 24th, I was cultivating some of the soybeans in the late afternoon and saw a big storm move through the area about five miles to the east of where I was working. I had two hundred acres of soybeans over in that area on land that belonged to my wife's uncle. About an hour after the storm had past, my wife was in the car sitting on the headlands of the field I was cultivating. She had been at her uncle's farm when the storm went through. The car had about two thousand dents on the body from some very large hail stones. She also indicated that the knee high soybeans on the farm were all gone. I stopped the tractor and rode back with her to see the damage.

I had never seen hail damage like this in my seventeen years with Gorman. What was a two foot plant before was nothing more than a little stick about ten inches tall without leaves. I did have insurance on the crop and the insurance adjusters showed up about three days later. Lucky for me, the neighbor farmer came over to me and said to take a full payout of $375 per acre with the stipulation I could still harvest if anything could be salvaged in the Fall. They agreed to let me do that, and after about two weeks you could not tell that the field was ever hit by a hail storm. The plants filled out with leaves, and I was able to harvest over thirty bushels of soybeans that Fall. The combination of $375 per acre from the insurance and over $250 from the crop, turned out to be the best soybeans I ever grew in my life thanks to the knowledge of my neighbor. The car was a different story with a thousand deductible and over $3500 in damage.

Campbell's checks were always good and always on time from the very beginning. It was a diferent story for the small family operations that we started supplying in 1973. We were forever looking to get paid, sometimes five or six months after we delivered a batch of compost to their farm. The mushroom cycle last about ninety days and the grower has sold the crop and has is money. They started to use Gorman like a bank and it got even worse when Compost Product's took over. In 1986, about five of the small family farms got together and sued us for poor compost and wanted their outstanding debt wiped off the books. Of course, we were sending the same compost, the same formula, to Prince Crossing everyday with very good results. The total amount owed was in excess of two hundred thousand and thankfully a judge seen through their defense and sided with our company. Most of the operations went bunkrupt and we never recouped a dime of what was owed, and stopped making compost for the small ventures altogether in 1987.

From the mid eighties to the late nineties, our old farm we were supplying, the oldest of the eight farms Campbell owned, out performed all the others on the most important number, the least cost to grow a pound of mushrooms. We produced compost over forty miles away, had the oldest facility, and still the cost to grow per pound was the least of all eight. We did make good compost, but it was the team effort and dedication from all involved that made the difference. We were always the envy of other operations around the country and hosted many visits from other mushroom companies even as far away as Australia and France.

Because the compost was so good, there was even talk in the early nineties of setting up a huge composting operation, run by our company, to supply compost to four farms from one central location in the Midwest. I was asked to put some numbers together and find a location but nothing ever transpired from it. We found that a site near South Bend, Indiana was central enough to supply Glenn, Michigan; Brighton, Indiana; Jackson, Ohio; and are old farm at West Chicago, Illinois and with economy of size, would have worked out very well, but it never went any further than that.

Fred originally purchased eighty acres in LaSalle County of which eight was used for the compost site. In 1985, an additional eighty adjoining us came up for sale. It was owned by fourteen different family members inherited from a family estate. Fred was approached by two individuals that said they represented all that were involved in the farm and wanted $1750 per acre. Fred never batted an eye and they all shoke on the deal an agreed to close as soon as possible. Over a month went by and the two men came back to the farm and said, "the rest of the owners were not satisfied with the price, and wanted more money." I thought the roof of the office was going to come off. Fred screamed and yelled for fifteen minutes saying they shook on the deal and finally told them, "to stick the farm up their ass and never come back." About a month went by and I saw one of the other owners in a little café in the town of Newark, Illinois. He asked why the sale did not go through? It seems everybody was fine with the original offer but these two individuals handling the deal were wanting more so they could pocket the difference and the other twelve would have never known it. About two weeks later, Fred bought the farm for $1500 per acre. Land in that area in 2009 is going for $8000 per acre without any improvements.

In January of 1987, I stopped farming all my land and sold all my equipment after four very hard years after Gorman had died. I had to pay the government over forty thousand dollars to get out of farming that year because of depreciation recapture. Fred figured I would now have some free time and was thinking about putting his now 160 acre farm into alfalfa hay. He did not like the fact that most of the time his trucks were returning to Chicago without something on board. We could haul bales of hay and sell them at the racetrack for additional income. We were also very efficient making compost, and most of the time we were done with our job for the day by one thirty in the afternoon. So, I had free time, I had plenty of help, if I had any bad hay we could sell it to Campbell Soup, and we had trucks going to the racetrack empty.

We decided to plow up the entire farm after the wheat harvest in the summer of 1987 and seeded it all to alfalfa in August. It came up beautiful. A full line of hay equipment, some new, some used, was purchased from the local John Deere dealer. We also purchased a special machine made by New Holland that picked up the bales off the ground after the baler had made them, stacked them hydraulically on a platform pulled by a small tractor, and then I could run this load of eighty four bales to the shed for storage without ever touching a single bale. No more back breaking work from stacking all that hay like I did as a kid. All looked like it would work on paper, but Fred would never see a single bale ever made or sold off that farm.

I got word in late September that Fred was very sick. I had not seen him for more than a month and was wondering what was wrong. In the prior four years, he always got to the farm at least once per week to see how is investment was doing. He was on the road a lot because he had a lot of investments. He would always stop at Gorman's when he supplied us with stable bedding back in the late seventies at least once per month just to see how we were doing. I was told in October that he had gotten pancreatic cancer and he died in early December of 1987. He left his wife, two daughters, and two son-in-laws to continue what he had built. Gorman was only fifty four when he died, Fred, I think, was only fifty nine. We did make hay the following year for the first time, but buyers came to the farm to purchase the hay and not a single bale was ever trucked back to the racetracks like Fred had envisioned.

Things transitioned very well after Fred died. I was very worried because of my experience with the two ladies after Gorman died. It was determined the day after Fred died that my superiors would now be the two son-in-laws. The following Spring my contract was reopened because of the extra work from the hay enterprise, and most importantly, Campbell's continued to receive good compost and they were kept happy.

Prior to Fred's death, I had approached him with an offer to buy Compost Product's. He turned me down saying that he was making way too much money. I figured, despite all the money spent on the new facility, he probably paid for the place in less than three years. It certainly didn't hurt that we used that old Jim Frezzo formula about pricing compost I learned in 1973. About a year after Fred died, I approached his wife with the same offer, but she and her family turned me down for a second time but pledged to give me first chance if it ever came up for sale.

The hay experiment worked well the first year, but I got less and less help the years after that. There was still a lot of hot, dirty work to do despite all the automation and machinery. I started working a lot of late hours, a lot of holidays to get the job done.

We also experienced a couple of rough winters on the hay stand and some of it was lost. It became very expensive to try and replant. The owners rented out any areas that was destroyed from the harsh winters and by 1996, the entire 160 acres was rented out to a local farmer to grow corn or soybeans. The hay project was over. We cleaned all the haying equipment up and sold it all at a local consignment sale for more than we had paid eight years earlier. Whenever I have free time now, I still help local farmers get the crop planted and harvested to help with my addiction to farming.

There has always been a tremendous amount of change in the Campbell mushroom division. Upper management, middle management, growers moving from one farm to another. Farms would grow mushrooms like crazy for a few months, and then grow nothing next to nothing for a while. Mushrooms like it cool and damp and probably why most farms are situated in the northern parts of the U.S. Campbell's took two big gambles to capture the southern markets by building a farm at Hillsboro, Texas, about fifty miles south of Dallas, and another farm at Dublin, Georgia, about one hundred and twenty miles south east of Atlanta, Georgia.

The Texas farm experienced cost overruns just on the building phase of over sixteen million. The cost of keeping these facilities air conditioned was enormous. Mushrooms grow best at about sixty five degrees and when the temperature is over one hundred for weeks on end, the bills were outrageous. The Texas farm sat about six hundred miles from the nearest stable bedding, so a formula of one hundred percent wheat straw was used. The straw was very cheap, but the amount of supplementation to get the compost up to a two percent nitrogen formula for good mushroom growth was very costly. We supplied our old Prince Crossing farm at twenty dollars per yard and delivered it forty miles. The cost of Hillsboro compost averaged over thirty dollars per yard produced just one hundred yards from the mushroom houses.

There was also a lot of disease and virus mainly due to the prolong high temperatures in the area and once it got started, it was very hard to stop. On one particular prolong event, the houses developed some unknown virus and the Campbell research facility at Nepolean, Ohio asked the farm manager to send up some samples for them to inspect and test. The farm manager at the time, Bill Edholm, could not send any mushroom samples for testing because there was not a single mushroom on the farm which should have been producing in excess of two hundred thousand pounds per week! The whole farm was eventually emptied out, sterilized, and had to start the growing process all over again. Some of the compost help from Texas wanted to visit our farm to see how we were making compost, and we had to turn them down afraid of getting that virus at our farm. In the ten plus years it was open, the farm probably averaged only seventy five thousand pounds of mushrooms in any given week, far below the two hundred thousand it was projected to raise.

The Dublin farm was also built in an area with no stable bedding within five hundred miles. They to used a straw formula that turned out to be very expensive and not very productive. The farm was built as an exact twin to a Dutch farm built in the Netherlands. All the equipment was produced overseas and created a lot of problems when something had to be replaced. All the compost was pasteurized and spawned by machinery and automation. All the houses were filled with machines, no human labor involved in anything. The biggest gamble was the use of mechanical harvesting, a million dollar gamble that never paid off. They did produce some good crops over the ten years they produced mushrooms, but was never very profitable. The Campbell experiment, in my estimation, was nothing but a big drain on the other six farms and the bottom line for the whole division for many years.

Large corporations can only absorb this type of set back for only so long. Campbell's made a decision in January of 2000 to sell the entire mushroom division, the eight farms, the research facility, all the spawn plants, all the vendor contracts, and the corporation headquarters to Money's Mushroom out of British Columbia, Canada. Money's was a well established mushroom company looking to get very large in a hurry. Some of the first things that occurred was to close and sell the two southern plants to a large mushroom conglomerate from the east coast. They agreed never to grow mushrooms from these facilites ever again hoping that would help bolster the mushroom market across the entire U.S. The less mushrooms on the market, the higher the retail price should go. The second thing, and the most costly, was to get rid of all the top money earners at all the farms, the people that new how to grow mushrooms. This would come back to haunt them sooner than later. I lost a lot of good friends at Prince Crossing when Money's took over. The quality of production was quickly compromised and much of the production had to be sold for soup production at about half the price of the fresh market. Volume also strated to deteriorate because a lot of the underlings were promoted and reached their level of incompetence. They even tried to cut our prices for stable bedding and for finished compost delivered to Prince Crossing. Money's did succeed in doing both by declaring bankruptcy in October of 2000, just nine months after they purchased the division.

Some of the first decisions after declaring bankruptcy was to close the farm in Jackson, Ohio and our farm in West Chicago, Illinois. The land our old mushroom farm we supplied was sitting on was valued at more than one hundred thousand per acre. They closed the farm up, tore all the buildings down, and put up half million dollar homes in its place. This sure made the town of West Chicago happy to have this, as they called it, eyesore gone forever and all the problems it created in its forty three year history. They never gave the 275 people that lost their job any thought or consideration in the whole deal.

We at Compost Product's were faced with a bevy of problems. We had an outstanding balance of $335,000 that was owed for stable bedding and compost production. I did not realize it had gotten that bad and might have explained why the owners were always getting mad anytime Money's name was mentioned in the final few months. We had two thousand tons of mushroom compost in rick form in all different stages of maturity that now had no home. We had about twenty five thousand yards of fresh stable bedding on the property waiting to be composted. Worst of all, we had all the contracts for all the racetracks in Chicago with home for only about half of what was cleaned up every day. We quickly had to switch gears and start thinking like Gorman or Fred would have done if they were faced with this situation.

Back in the early eighties, Fred convinced the old Ralston Purina farm to clean the garbage out of the old compost from the mushroom houses so he could sell it back into the Chicago market as mulch or soil amendment. His trucks were there dumping fresh stable bedding and had to go back to Chicago anyway. They agreed to place a Trommel screen unit in the emptying process that would remove all the junk and produce clean spent mushroom compost. He moved a lot of material over the years. I convinced the owners that we could make a product similar to spent compost with the advantage of a lot less salt and a lot less smell. We would not have to use chicken manure or potash that has a high salt content and one of the big disadvantages with spent mushroom compost.

It took about six months for Money's to come get their equipment off the farm. We continued to use the machinery for turning until it would not form a rick anymore because of the shortness of the material. We then ran the product through a large manure spreader for a couple of weeks and piled. We finished off the two thousand tons in rick form and then started to chew on the old stockpile of stable bedding mixing it fifty-fifty with the new material coming in from the tracks. We did not have the big fire of all that old material like the O'Reilly ladies had some seventeen years earlier to our great fortune.

By closing the Prince Crossing farm, it left a void of about 1200 yards of spent compost per week in the Chicago market that was being used prior to the bankruptcy. Compost Product's was going to try and fill that void with our own version of spent compost. We still had the problem with all the garbage and something had to be done about that before we could sell it to the public.

The owner's of Compost Product's Inc. did receive about sixty five cents on the dollar from the bankruptcy after about a two year wait. That check along with free use of their machinery for six months, all that material in rick form, all that material in stockpile, and we probably came out ahead in the whole deal and our mushroom days were behind us for good. With the loss of all that turning machinery and the need to clean the garbage out of the material, we had to do some equipment shopping very soon.

Life After The Mushroom

In the winter of 2001, I became acquainted with Les Kulhman and his wife Joan from northeastern Colorado. He had over thirty years of experience in the compost industry as a hands on operator and consultant. He was one of the first to design and build a self propelled straddle turner for windrow compost production. Les was one of the first innovators in the development of windrow composting. Long, wide rows of material to compost beef and sheep manure out in the Colorado feedlots. They were extremely crude looking pieces of equipment when it was first tried. It sat on two large wheels in front, two smaller caster wheels in back and wide enough to straddle a twelve or fourteen foot windrow. A large engine sat on top of a platform eight to nine feet high with a cab for the operator. A clutch was attached to the engine and when it was engaged, would power large belts that ran along the edge of the machine down to a big drum resembling a roto-tiller under the platform running very close to the ground. As you move forward over the windrow, the large drum that ran at speeds of four hundred rpm's would physically pick up the material and vigorously eject it towards the rear of the machine forming a new windrow behind. The composting process would occur with the help of all the oxygen inducted into the material as you moved along the windrow. Everytime through the pile with the turner the material got shorter, and after a while it would resemble dirt with some structure. He called his new invention a "K/W" or King of the Windrow turner.

With the advent of the computer, I began looking for a self propelled turner to replace all the machinery Money's was coming for someday. I found one of these "K/W" turners that Les had built out near DesMoines, Iowa that he wanted to sell. It was used on a farm near there that had gone bankrupt a year earlier. It was capable of turning a windrow that was sixteen feet wide and about six feet in height. From my calculations of what we had coming in and the amount of usable concrete space we had at any one time, this seemed like the perfect fit. I let the owners know about this machine and a week later we made the four hundred mile trip to see and run the thing for ourselves. Les indicated he wanted $65,000 for the turner as is and where it sat. Jim, one of the owners, and I left early one morning to see the machine that sat at a machine shop in Winterset, Iowa, about forty miles southwest of DesMoines. It was five below zero when we arrived, but the engine was plugged in for warmth and we started the machine right up and drove it around a little. We headed back to our home base in LaSalle County and decided the machine was priced a little high for the condition it was in.

Again with the computer, I was able to find a lot of these machines around but none as close as this one. Back in 1990, Illinois and a lot of other states banned yardwaste from landfills. Huge operations emerged over night trying to make a buck composting all of this new yardwaste material. Most sites were built close to the yardwaste source to eliminate trucking costs and unfortunately very close to many communities. A lot of activity went on before rules and regulations were in existence and many lawsuits were filed for odor production and nuisance. The cart before the horse senerio. The old rule of thumb has always been, if you can see a house, your too close to compost. Odors will travel up to a mile in extreme cases, but will normally disapate in less than 1500 feet. Hundreds of operations sprang up with odor production never considered mainly due to formula ignorance. There were a lot of these turners produced and sold to some of these companies that went out of business in less than three years. Most did not know how to compost or what to do with it once it was made. A lot of litigation the first few years and a lot of equipment came back on the market.

When we got back from the Iowa trip, I informed Les that we were going to look a little further for our turning needs. He called back the same day and said, "he could deliver the machine and give us one day training for the price of $52,000". The owners accepted his offer and we were formally introduced to Les about five days later. We started to make composted landscape mulch for the greater Chicago area market by the windrow method in February of 2001.

We started to learn a new way to make compost once we got the K/W. A whole new learning curve yet we were using the same material when we made mushroom compost. At first, we used way too much water and the material turned into mud because the K/W was so aggressive. With mushroom compost, you bring the moisture up to 75% as fast as possible, and hold it there for the two week process. Mushroom equipment is very gentle whenever it touched the compost, where as the windrow turning equipment is extremely vigorous to the point of being violent.

One of the toughest jobs was trying to put water on mushroom compost when it is below zero. Everything had to be orchestrated ahead of time so the hoses would not freeze during the day. Everything would have to wrapped up at night and placed in a heated shed so they would not freeze over night. Everything we did in the winter took twice as long to do than in the summer time. With the windrow composting system, we put a little water on to begin with, and that was it. That made the Winters much more bearable.

So we backed off on the volume of water very quickly since it was going to be on the concrete for two months. A lot of time for mother nature to have some effects. Every time the K/W went through the windrow and oxygenated the material, the length of the material became shorter. The smaller the pieces, the more water it can absorb and hold. There is less of a chance for evaporation which is why compost is so good around plants for water conservation.

Life after mushroom compost was almost seemless. The company was cleaning and hauling out all the stable bedding from the Chicago racetracks, they delivered three to four thousand yards per week to the old Ralston Purina farm, which had gone through bankruptcy itself and was now called Terry's Food from Minnesota. All the excess stable bedding was delivered to me to process into compost that looked similar to spent mushroom compost. We continued to remove fifteen to twenty loads per week of spent from Princeton back to Chicago or to me to stockpile for future delivery. The entire Chicago market probably doubled the amount of compost it was now using over just five years earlier. Several other organic materials became available that people were willing to pay us to take them off their hands. We accepted wallboard, old or damaged hay and straw, yardwaste, leaves, woodchips, paper, cardboard, spoiled grain, road kill, pre- and post-consumer food waste, sod, used lumber and construction debris, and contaminated soil. My goal was to blend any of this to an initial C/N ratio of 35 to 1.

The C/N or carbon/nitrogen ratio is the total amount of carbon in a given product compared to the total amount of nitrogen in a product. The lower the C/N, the faster the material will compost, and the more it will shrink. It will also produce more odor in the composting process. The higher the C/N, the slower it will compost, the less it will shrink, and the less odor will be produced. The C/N ratio, or the lack of knowledge of it, was the main reason so many operations failed when the yardwaste ban took effect in 1990. The closer you can get to 35/1 C/N, the better the operation will be for everything involved, including neighbors.

Having the proper C/N ratio to start and turning with the K/W turner reduced odor emissions to the point of almost being non-existent. I was hoping to learn from all those earlier operations, and not to repeat their mistakes here. With mushroom compost, the only thing you could do was to turn the compost on a regular basis. Mushroom compost starts at 14/1 C/N ratio and only gets worse until you deliver and fill two weeks later.

We also developed a new business by separating out all the large woody materials when delivered. We would hire a custom grinder every three months or so to come in and grind all this material up to less than three inches, put a little compost water on the material to aide in darkening, and about six months later sell wood mulch into the Chicago market. The ideal situation by getting paid for the inbound material, and getting paid for the outbound material.

We used the K/W turner from February 2001 until October of 2004. The only time I had any problems with the ten year old machine was on September 11, 2001. For some reason, the machine would not move forward or backwards. Starting at nine in the morning, one of my loader operators came over and said, "a plane had hit one of the towers in New York City." I did not give it much thought since I was trying to figure out what my problem was with the machine. About an hour later, the same operator came over again and said, "the other tower was hit, a plane hit the pentagon, and a plane had crashed in Pennsylvania." I still could not really grasp the magnitude of what he was saying still trying to get the turner to move. Just before noon he came over and told me, "that both towers were now flat on the ground." I stopped what I was doing and what the other four men were doing and went into the office. I found an old black and white television, plugged it in, and seen first hand all the destruction that had happened that morning on the east coast. The President was in the South at the time and was flown out into the middle of Kansas, I believe, until it was safe for him to return to Washington.

The skies over the compost site were always filled with white jet streams from all the aircraft that flew over our area. The FAA had grounded all flights around ten in the morning, and the skies were perfectly blue from horizon to horizon on September 11th.

Around four in the afternoon, I did see a very strange site in the southern skies about twenty miles from our compost site. There was a large blue and white 747 escorted by two small fighter jets all flying at about fifteen thousand feet heading east bound very quickly. I figured it must have been President Bush heading back to Washington. I was standing on top of the K/W turner at that time, and after watching the planes go out of sight, I looked down at my machine and noticed a wire was off on the backside of the engine that was used for grounding purposes. I reattached the wire, started up the machine, moved it out of the way, and went home to be with my family.

They say you never forget certain things and what you were doing at the time. When I was in my seventh grade gym class close to Thanksgiving, the gym instructor, a rather large, ominous looking fellow, walked out of his office and onto the gym floor crying. He told us President Kennedy had been shot in Dallas, Texas. I was standing in the little farm office talking to Gorman at 8:15 on May 11, 1982 when he did not answer with his usual 10/4 over the FM radios. A half hour later we found out he had died at that precise moment. I got home from my junior year at the University of Illinois on June 3rd, Mom and I went up to stay with Dad in the hospital. We stayed by his side for the next five days and five nights. We could no longer take it, so we went home to get some rest only to be called back to the hospital shortly after we got to the house. Dad had turned for the worse, and Mom and I just made it back to his hospital room before he died. I was working on the K/W turner trying to make it move when our country was attacked on September 11, 2001. Over three thousand people killed and I was worried about turning some compost. That's how an addiction works. It finally dawned on me that there was something more important than composting or farming at that particular moment.

Because of our ability to use and accept more raw products for composting, we went shopping for a larger compost turner in

March of 2004. I had attended a compost conference in Columbus, Ohio in the winter of 2004. Of course everything connected with compost is advertised at these events. One particular vendor was represented by a short, very energetic, full of fire gentleman of about 75 years old. He was one of the salesman for Wildcat Manufacturing Company out of Freeman, South Dakota. The show kicked off at seven in the morning and ran to about ten each night for three days straight. This man never stopped talking about all the attributes of this turner for all three days. I never seen anybody with this much energy at that age. I needed to look into Wildcat for our future turning needs.

I contacted the main salesman, Tim O'Hara in Freeman. Literature was sent along with some pricing and we pondered the purchase for about five months. I asked Tim if I could go see a machine similar to what we were think about buying that was in service. He indicated there was a machine in Wichita, Kansas sold about a year earlier. Wildcat contacted this farm for me and a date was set for me to fly to Wichita, drive the thirty miles to the compost site, and fly back that evening. I was able to purchase tickets with Southwest for $29 each way, I paid $25 for the rental car, and got to see and run a machine very similar to what we were thinking about buying for the entire day. I was sold as soon as I laid eyes on the machine. The K/W was certainly impressive, but nothing could compare to this machine. When I returned from my trip, I asked the owners if I could negotiate for the turner, and they all agreed.

I had learned from the best about price negotiations. Gorman was forever selling high and buying low long before I met him. Early in my career with him, he made a deal for two new Chevy semi trucks, two new Chevy pickup trucks, and two new Chevy grain trucks all over the phone from a dealer in Garden City, Kansas. They put the pickups inside the grain boxes, and the four man carvan was to arrive on a Tuesday afternoon. They did not show until Friday morning. It seems they hooked up with a bunch of women in DesMoines and brought them along for the ride. Anyway, it was an unbelievable deal with them taking four used vehicles and all of those women back with them. I heard that about a year later this dealer was out of business.

The original price quoted for the turner was $236,000 delivered to Compost Product's in LaSalle County, Illinois. After going back and forth with the salesman until October, the final price was $196,000 delivered and included an extra set of cutting edges valued at $4,000 that need to be replaced about every two years because of wear. It had double the turning capacity of the K/W and we were able to put more material on the concrete because we could now make the windrows eight feet tall and eighteen feet wide. When we made and turned mushroom compost, it was good to handle about two hundred tons per hour. The new Wildcat could handle in excess of six thousand tons per hour! That's right, six thousand tons per hour! I was fortunate to fly out and inspect the machine in Freeman before it was delivered. I flew into Omaha, Nebraska and drove about 130 miles north and saw this amazing country for myself. I absolutely fell in love with that farming area and if I become a millionaire from this book or the lottery, I will buy a lot of land around Freeman. I saw a speed limit sign in South Dakota near Freeman that read, "Speed Limit 75 or whatever feels comfortable." It's quite the place to see.

We had two windrow turners for a while. Like we did years before when we cleaned up all the hay equipment and sold it for more than we had paid, or when we cleaned up the old tub grinder at Gorman's, and sold it for more than we had paid, we cleaned up the K/W and advertised it for more than we paid in some of the compost publications. I received a call from Les Kulhman, the original owner, in late November to see if the turner was available? He wanted to come see it that afternoon even though he was calling from near Denver. He arrived at our farm at about 3:00 that afternoon, drove the machine through a few small windrows, and decided to give us $55,000 for the machine. It ran trouble free except for September 11, 2001, the better part of four years, and we got all our money back. To this day, I never hesitate to call Les if I have a question and he and his wife can certainly be added to my growing list of most respected.

Me at age 14 driving what I considered at that time a large International tractor and a five bottom plow for one of the local farmers after school in the Spring of 1965.

My picture after receiving the FFA Star State Farmer Award in Springfield, Illinois in the Spring of 1969. Less than 1% receive this award in any given Year. A lot of thanks to my FFA advisor Ron Deininger.

The only picture I have of my wonderful high school Agriculture teacher and four year advisor in FFA, Ron Deininger in 1969.

Very early picture of Gorman and his hay making crew before I went to work for him. He is the man all the way to the left. This was well before a lot of mechanization for handling hay. Just a lot of plain back breaking work.

Another picture of Gorman standing by a nice field of Popcorn in the Arsenal probably in the Fall of 1974.

Some of the round bales waiting to be ground by the "WHO" tub grinder down in the Arsenal about ten miles south of Gorman's home base. Not unusual to grind a bale every minute with that amazing machine but never pleasing Campbell Soup at any time from 1970 to 1974. The round baler made by Vermeer Mfg.

One of the earliest pictures of making compost for a few small mushroom growers with the old choke feed turner that we purchased off of Campbell Soup in 1973 and the Haymaster wheel loader we purchased to load all of that ground hay after our initial loader caught fire in the Arsenal. A second HayMaster can be seen in the background by the white corn crib.

December 7, 1984, a sad day for all involved. Photo of Gorman's auction of all his farm equipment. Picture only shows about half of what was sold that cold day by his wife and daughter. A huge crowd at his auction but not at his funeral.

Jim Frezzo's compost site in Avondale, Pennsylvania on April 10, 1973. Gorman sent me out to see what I could learn about composting and prices. Jim and his family took me in like I was one of their sons and showed me everything they knew. No shortage of water to get the materials wet at his site.

A picture of me and Bob Pannell standing next to one of his infamous Cross-Mix Turners on my initial trip to Kennett Square, Pennsylvania in 1973. He was able to stop laughing long enough to take this shot after I had just gotten thrown off another compost site trying to learn about composting before I visited Bob.

Mike Pia shows how you can really find out if it is good compost by taking a large handful and squeezing it and smelling it. This was again on my initial trip to the compost capital of the world to learn about composting and pricing in 1973. Site was in Kaolin, Pennsylvania.

Fred Noorlag who gave me a second chance to compost and farm with his new company, Compost Product's, Inc. in 1983. This was taken in 1985 working on the irrigation system for the composting process.

What mushroom composting looks like with a large, modern choke feed turner, a new John Deere Payloader, building the initial rick so the self-propelled turners can take over. Material is already over 160 degrees and about 75% moisture, ideal for the fourteen day process to begin.

A self propelled Pannell Compost Turner turning a rick of compost in 1996. After the initial build with the choke feed turner, this machine would run down the ricks five times in a fourteen day period before being shipped and filled at Prince Crossing farm some forty miles away in West Chicago, Il.

Using the supplement spreader to pull the tarp roller over the ricks after they are turned and cleaned up for the day. The supplement spreader was filled with nitrogen supplements and used to meter them on top of the ricks before turning to bring the material up to 2.5% nitrogen for good mushroom growth. Tarps were used to hold moisture in during dry times and the cold windchill out during the Winter months. Each tarp is 200 feet long and 18 feet wide.

A lot of stable bedding being watered and piled to start the compost process. A very simple process before Money's talked us into using a pre-wet machine. You can also see the very nice supplement storage building in the background that held all the different nitrogen supplements.

The Power Screen Compost Trommel that all the finished compost goes through before being sold to our customers to eliminate all the garbage. Started to use this after we stopped making mushroom compost in 2000.

Picture of our first load of hay off the farm in 1988 and one of the fleet of trucks that would haul compost off our farm to Campbell Soup and remove stable bedding from the racetracks in Chicago.

Me and Tom Creech in front of a pile of stable bedding or as he calls it, muck, in Lexington, Kentucky at his beautiful bale factory in 1990.

Creech's bale factory with finished one ton bales coming out for shipment to mushroom farms all over the country. Not unusual to make 200 bales per day!

Machinery, Innovation, And Imagination

I probably have gotten far enough along in this book without mentioning the fact that the mushroom composter does not get enough credit for all the advances we take for granted and that has happened over the last two hundred years. All the trials and tribulations have never been accounted for by anyone in the present composting industry to my knowledge. Yes, composting has gone on forever, but it was the mushroom composter that determined what could and could not be composted. It was the mushroom composter, with a lot of back breaking work, that first realized how to prepare a special product so mushrooms would thrive and nothing else. They were the first to uncover the mystery of sterlization when the compost got over 160 degrees and what we use today as a benchmark for killing deadly pathogens in waste treatment plants. He also uncovered how things shrink by more than half when proper composting takes place. A lot of the composted landscape mulch we make today will shrink by 90% after a year and is the real meaning of the carbon cycle, returning organic matter back to its original state.

When I first got involved with compost, there were yearly gatherings that dealt with the composting industry other than those involved with mushroom composting. At first, these annual meetings would draw about seventy five people or less for one single day. The last national show I attended lasted three days, drew over one thousand participants, and had over one hundred vendors that sell to the compost industry. No one ever mentions the old mushroom composter. When all that yardwaste started to get diverted from landfills, it was the knowledge of the mushroom composter that everyone fell back on for information. They already knew about proper C/N, shrink, moisture, temperatures, and odors. Maybe this book will stir up enough people in the right places to give the old mushroom man his due.

A short review of mushroom compost before we move on. The main ingredient has always been stable bedding, but you can use wheat straw, hay, or corn cobs. You lay out a known quantity of raw feedstock, add some water and some nitrogen source, usually chicken manure, pile the material for the heating process to begin. Once a good deal of heat has developed , special machinery is used to place the material in ricks, where the Thermophilic bacteria take over and start to eat on all the carbohydrate in the feedstock. Temperatures soar to over 160 degrees, oxygen becomes depleted, and the material is mechanically turned to add water and oxygen. All this sounds simple until you introduce people, machinery, and mother nature.

Compost can be made in an aerobic state or an anerobic state. Mushrooms and neighbors like it best when it is in the aerobic state. When we worked for Campbell's, despite the weather, manpower, or machinery, we ran such tight control on the finished product delivered to Prince Crossing, we were given the prestigious award of "Select Supplier" to Campbell Soup in 1988 and every year after that until 1999. Targets or goals were set up for us to try and achieve on a everyday basis no matter what conditions we encountered. Some of the goals were; weight delivered, 210,000 to 220,000 pounds/day, moisture, 72% plus or minus 1%, PH, 7.9 to 8.1, nitrogen content, 2.3 to 2.5, color, dark brown to black, length, 4 to 5 inches, fourteen day temperatue record, 160 plus all two weeks, and about a dozen other things we kept track of everyday.

There was a big deal and fan fare made of this award for the fact we were only the second vendor of all the over 500 vendors Campbell Soup had to receive the award. Frank Perdue Farms was the first for delivering quality chicken to the soup plants. Dr. Bob Miller was in charge of the entire mushroom division for Campbell's and made a special trip out to LaSalle County to present us with a plaque and flag. He personally singled me out and said, "that I reminded him of an old country and western song by Barbara Mandrel, I Was Country When Country Wasn't Cool, and that I was certainly Compost When Compost Wasn't Cool." What a great honor for the farm and for me and I never forgot that day and the reason the book is titled like it is.

The actual process to grow mushrooms would fill a small library. If there is an urgent need to learn how to grow, I would refer you to Penn State University which is the capital of the world for mushroom research and development. Needless to say, no matter how good the compost was the day it was delivered and filled, a lot can happen inside those special growing rooms over the next ninety days. In the real estate world, everybody talks about location, location, location. The three most important thing to a mushroom grower is compost, compost, compost. Yields will vary by 100% at any given time without warning, but the compost always gets the blame but never the credit.

When we started to make the composted landscape mulch, we took away all the nitrogen supplements, we took away a lot of the water, we added about a year to the process, but you can never take away the need for oxygen. We went from loaf like ricks that were eight feet wide and seven feet tall, to windrows that were eighteen feet wide and eight feet tall. When we had more material than would fit on the concrete pad, we would make windrows out in the field on dirt. It was a little tougher to work out there when it got wet , but we had a lot of material coming in that people were paying us to get rid of and I did not turn anything down. We ran a very intense program of five or six weeks on the concrete pad turning the material three times per week. When the material got good and black, and the temperatures stayed below 150 degrees, we would take the large pay loaders and pile the material off the concrete on the dirt for what we called the "Curing Phase". By this time, 90% of the shrink had occurred from the original volume, and 95% percent of the smell was gone. We would roll this material that was curing three or four times over the next six months with the large pay loaders to add some oxygen and insure there was no odor production.

The one big problem that needed to addressed was the garbage in all the raw feedstock's. Princeton Farms cured this problem with the help of Fred Noorlag many years ago in order to have a product you could sell. They installed a Power Screen Compost Trommel at their farm and cleaned up the old compost so it could be sold into the Chicago market. We would have to do something very similar if we wanted to sell also.

In my forty years of looking at stable bedding and yardwaste, I have never found a bag of money. I have found bed mattress's, car tires and rims, batteries, hubcaps, beer cans and bottles, twines off straw and hay bales, old feed, salt blocks, and the very worst thing, plastic. You may remember the movie , "The Graduate", where somebody tells Dustin Hoffman the future was in plastic. I felt like I was in the movie on a seven day per week basis when I got the addiction to compost. It was in your best interest to try and pull out plastic before entering the turning phase of your compost feedstock. When we started to use the windrow turner's in 2001, it was even more important than ever. When the turner went down the windrow, it would take a piece of plastic and make two pieces out of it. The next time down, it would take those two and make four. Everytime through there was more pieces that were getting smaller each time and very hard to get out of the finished material. Go through the pile fifteen times, and you get the idea.

We made a decision to purchase a compost screener in the summer of 2001. This very large, expensive piece of equipment had a hopper about twelve feet high that was filled with a pay loader of finished compost for cleaning. Once dumped in the hopper, the compost would move laterally into a large six foot diameter drum that would rotate at about twenty rpm's. The walls of the drum were filled with three quarter inch holes so only the compost could drop through and would allow the garbage to roll out the back to be disposed of. The trick with all compost screeners is to have the material as dry as possible so all those holes do not fill up with wet compost. There is a show on TV that has a man doing different dirty jobs. He may want to try and clean the interior of this drum and all the little holes when the compost is too wet and it's a hundred degrees outside for one of his shows.

When we made the decision to purchase, I had two different vendors show up the same day to demostrate their machines. It was a planned turn of events so we could judge two of the best machines in the business at the same time and for the best price. We parked one machine, a ten year old Power Screen, on the north end of the concrete and a pay loader feeding the screen. We parked the other, a brand new rig with the name withheld on the south end with a pay loader feeding that one with finished compost. For the entire two hours we screened both saleman were running between the two screens and calling their boss to see what price they could offer to get us to buy. It was an absolutely perfect setup. We ended up buying the ten year old unit over a brand new unit for about $25,000 less than was quoted when we first started to use the machines that morning. About $12,000 per hour savings I guess you could say! Since that day, everything produced on the farm goes through that Power Screen Trommel machine before it is sold.

During the early part of 2006, we started taking in much more yardwaste. We were accepting a combination of grass, leaves, limbs, large tree branches, and anything the homeowner could put into a recycle bag. We tripled the amount of clean drywall that particular year. We started off charging five dollars per yard for the wallboard, and quickly went up to eight dollars with no drop off in volume. This translated into about twenty dollars per ton when the local tip fees at the landfill were running in excess of fifty. It was a good business decision for the people dumping the wallboard and mixed with all that yardwaste, it made a beautiful compost. We used a formula of fifty percent stable bedding, forty percent yardwaste, and ten percent wallboard, composted it for two months, piled it for six months to cure, and started selling it in 2007.

I started to read about the benefits of gypsum wallboard, calcium sulfate to be exact, and corn production from Midwest universities. About the same time, fertilizer prices and corn prices started to rise substantially in the Fall of 2007. I penciled out different application rates of the Gypsum/Compost blend which I officially called "Gypost", to local farm fields in comparison to dry fertilizer rates that were normally used. Late in 2007, I applied twenty yards or ten tons per acre of Gypost on a twenty acre test plot on our own farm that was rented out to a local farmer. I still had the original 1987 John Deere tractor from our haying days, and I had a large Knight manure spreader purchased in 1988, a time when we had more stable bedding than Princeton or Campbell's needed and we spread stable bedding on the farm to get rid of it. I tested the Gypost for Nitrogen, Phosphate, and Potash and made a determination that ten tons per acre was equivalent to the normal amount applied for a 180 bushel corn crop. In addition to N,P,K, the farmer would also receive many micro-nutrients like Sulfur, Boron, Zinc, and of course a very healthy dose of Calcium. In addition to all that, organic matter would be supplied that no commercial dry fertilizer could provide and was hard to put a price on that.

The price of fertilizer to feed a 180 bushel corn crop that Fall would have cost the farmer about $335 per acre. Since we had about sixty five percent of our income already from the fees on the inbound material, I priced the Gypost, with the consent of the owners, at $125 per acre spread on the land. I was waiting for the test plot results before I started to advertise in 2008. The field was planted in the Spring of 2008, and was harvested in the Fall with yields that were significantly better than the rest of the field.

The field did suffer from a very bad case of rootworm that year except where I had spread the Gypost. There looked like some type of suppression of the rootworm in that area, thus there was more yield. I also seen, as the university research indicated, that the ground worked up a lot better and held much more water where Gypost was applied.

We did not have to spend a single dime on advertisement that Fall. The farmer that harvested the corn off that test plot hit the local coffee shops and started to talk about the benefits and price of Gypost. In less than two weeks, I had all the land I could possible spread with one tractor and that old spreader. I had to turn down more land than I spread that Fall because the owners would not go out and buy a second tractor and spreader. It was always my dream from the day we lost the contract with Money's mushroom, to take all the material, compost it, and spread it out on the farmland in LaSalle County and still make money from doing it. This is very common in California or Florida where they raise thousands of dollars of crops per acre. I never thought I would see it here in Illinois on corn or soybean fields. My dream had finally come true.

I spread Gypost, with very little help, on 770 acres that Fall and Winter of 2008 on fields very close the compost farm. I even spread on Christmas Eve and New Years Day. With all my excitement over spreading, I had completely forgotten one important aspect of me doing all this extra work. Nothing was negotiated before the spreading started. To my great surprise, I received nothing for all my extra labors, innovativeness, and salesmanship. I brought in an extra $96,000 to the farm that year with a twenty two year old tractor and a twenty one year old manure spreader, and at a cost of less the $8,000 to do all the spreading. I guess the material was better off sitting in a curing pile shrinking away than out on a field making money for the operation. I thought I was still working for Fred Noorlag, and he would have never treated anyone that way. It sure left a bad taste and a hint of the new management coming into the organization in the form of Grandchildren.

To cure the problem of no extra money from spreading Gypost on the local farm fields in 2008, I set up a payment plan of ten dollars per acre for every acre applied in 2009 for myself. I raised the price from $125 per acre to $130 per acre to cover half of the income I would receive. I thought I could sell all I could spread again in 2009, but the ownership said they needed at least $165 per acre to make a good profit. I have no idea how they came up with that number other than outright greed from the grandchildren. When I approached the farmers with the new number, not a single acre of Gypost was purchased or applied in the Fall of that year. It was a triple wammy, corn was down twenty five percent, fertilizer was down thirty three percent, and Gypost was increased thirty five percent. You may never see Gypost applied to farm fields again in my lifetime with that type of management. The dream was gone after just one year.

Because of the amount of wallboard and yardwaste being accepted in early 2007, I asked if we could buy our own grinder instead of hiring a custom grinder to come in what looked like every month or more? The government also made the decision a little easier with the announcement to give small business an extra $100,000 off the income tax on equipment purchased that year. I set out on the computer and found what I thought would be the correct size and style machine that would suit our needs for a long time. It was a Vermeer Horizontal Grinder with 2900 hours on it located near Holland, Michigan. The asking price as is and where it sat was $129,000.

Unlike the tub grinder we used years earlier at Gorman's farm where the material was dropped in from above the rotating hammers inside a large tub, this machine would be loaded from the side of the rotating hammers and fed horizontally into the mill for grinding. All the yardwaste entering the the farm would either be ground and composted or the large woody material would be ground to about three inches or less, and sold as mulch.

I set off to see this machine with the owners telling me, "if I liked it, go ahead and purchase the machine for them." About a week went by and I set up an appointment to see the machine in action up near Holland, Michigan. There was no way I was going to drive that far and not go ten miles out of my way and see the old Glenn mushroom farm that sat about thirty miles south of Holland. This farm, along with making compost for themselves and growing mushrooms, also produced compost for Prince Crossing for a time before we took it over at Gorman's back in 1978. Money's continued to operate this farm after they acquired it back in 2000 from Campbell's until about 2004 when it was closed for financial problems and the continuing bankruptcy litigation. A large dairy was built across the road from the old farm since the last time I visited the area. I learned that when Money's sold the property, over two hundred acres, the dairy bought the land and buildings. They farmed the land and used all the buildings to house young dairy animals.

The old place sure brought back a lot of memories. There was over three hundred people working here at the peak of production. Usually over three hundred thousand pounds of mushrooms were harvested and sold each week. The most vivid memory was the amount of snow they endured each year at this site. They were sitting in the worst spot on earth for lake effect snow off of Lake Michigan. I still have a feeling that somebody in a shirt and tie at Campbell Soup corporate offices decided it would be a great place to grow mushrooms. It was a remote location, odors would not be a problem, it was cooler in the summer and warmer in the winter because of the lake, but nobody figured in the snow that dumped on this site for days on end. People had trouble getting to work, materials had a hard time getting to and out of this place on many occasions. I think the person who made the decision had this God forsaken piece of land and wanted to get rid of it in the worst way. This is only my opinion, and never verified any truth to it.

This farm did do all the composting under roof. The snow was so bad in 1979 that the roof covering the compost area collapsed on top of all the ricks and turning equipment one evening when no one was working. It took about a week to clear all the wreakage and snow out of the area and repair the damage to the turning machinery. We thought we had it bad that year by cleaning all the alleys with shovels, but nothing compared to their mess that same miserable year.

I had no problem finding the site where the grinder was to be demonstrated. The machine owner said it sat next to a large landfill. I could see this mound of garbage from over fifteen miles away. Everything about the machine looked as the owner had described it over the phone. I did notice one very big problem. I had learned from Gorman years before that before you start any engine, check the oil. Before you buy anything, check the oil. That dipstick and that oil sample can tell you a lot about how the engine has been cared for before you buy. In my seventeen years with him, we never had an engine problem. He considered oil like blood in a human, it was the life line.

I climbed up on the platform to check the oil and noticed that the oil fill cap that leads directly to the oil pan on the engine was missing. It was replaced by an old rag that appeared to have been there for many hours of operation. I went ahead and checked the oil anyway and it showed that it sure needed to be changed and a lot of abuse had been done. The owner went ahead and ground anything and everything he loaded into the grinder. The end result was a beautiful looking product out the back and performed a lot faster than I would have thought for this size machine. The whole time he ground material I was thinking about that oil fill cap and what Gorman would have done.

I left after about an hour and told the owner I would be in touch with him after conferring with my boss. A Caterpillar engine with that horse power would probably cost over thirty thousand after installation. I told my boss that we should make an offer of $85,000 or start another search on the computer. Nothing like that size or price came up on the computer unless you wanted a new machine for $285,000. I went back and forth with price for about three weeks even having the local Vermeer dealer from Grand Rapids go and look at the machine to make sure

I did not miss something. It was delivered to our farm in LaSalle County for $92,000 about a month after I first looked at the machine in Holland, Michigan. Knowing how to negotiate and the importance of oil saved the farm about thirty seven thousand. Upon arrival, we changed the oil and greased all the zerks that had not been touched for a very long time, and have been using it about an hour everyday without any problems. I got the machine I wanted in the end and little did Mr. Vermeer know we had met again with the purchase of that grinder.

I do remember one engine problem that we had in my seventeen years working for Gorman. It was in the Spring of 1978 down in the Arsenal. One of our oldest employees, seventy five year old Henry DeGroot, was working ground up ahead of the planter. He called me on the radio saying the engine was leaking water and blowing up on the windshield of his large four wheel drive tractor. I told him to shut the tractor down right where he was. It took me about twenty minutes to get to him and discover that he had stopped in the middle of a turn and the tractor was cocked all the way to one side in order to make the turn. I got close to the engine and discovered a connecting rod that holds the piston inside the engine was now sticking outside the engine. I had never seen anything like this and either had Gorman.

We called John Deere to try and come get the tractor the following day. When the driver for John Deere arrived, he indicated to Gorman and I that there was a recall for this particular model, John Deere 8630, and his had not been called in for the service update as of yet and the reason the engine did what it did. The $12,000 that was needed to repair the tractor came out of John Deere and not a penny out of Gorman's pocket. What a relief, but our other problem was how were we going to get this machine straightened out for loading and the implement off the back of the tractor? The driver said, "get in it and start it up. You can't hurt it any worse than it already is." I climbed up in the cab, holding my breath. I turned it over, and the engine started right up. We unhooked the implement off the back, and drove it right on the flatbed. I guess you learn something everyday. A week later, the tractor was back in the field and the total cost was zero.

In the Winter of 2007, I attended the U.S. Compost Council seminar in Orlando, Florida. There was close to a thousand people in attendance. On the first day of meetings, I received a note that Joe Dinorsia wanting to see me. He worked for Laurel Valley Farms in Avondale, Pennsylvania. On my first trip to the mushroom capital of the world in 1973 to find out about composting and pricing, Laurel Valley was one the operations that would not let me set a foot on the property or ask a single question. I did not remember at the time I got his note that I had actually met Joe in 2000.

He came to Compost Product's to try and help out the farm we were supplying compost, Prince Crossing. It had been under Money's management for about four months and the mushroom houses started having serious fly problems. These things are not the normal house fly but a one tenth size version that deposit their eggs in the compost after it is filled into the house. When the eggs hatch, the larvae bore up the stem of the mushroom and suck out a lot of nutrition making the mushroom impossible to sell to the fresh market. It had gotten so bad inside the house's that the picker's had to wear "bee masks" in order to breath and see to pick anything.

Production started to plummet and they turned to Joe and his company for some answers. They sold a liquid product that could be applied to the compost on the last turn before it was delivered to the mushroom farm. It would sterilize the fly eggs and would not let them hatch into larvae. He came to Compost Product's to help us install a water tank and a small electric pump on the compost turner and teach us how to use the very valuable liquid. It did help reduce the fly population very quickly, but it was only about three months later Money's would declare bankruptcy.

I finally caught up with Joe on the second day of the seminar and all those old memories came back to me. He and another individual from Germany along with Laurel Valley were putting together a national group of blenders to blend and bag "Green Roof" material. Green Roof is a product being used to grow plants on top of buildings, mainly in large cities, and has become very popular in the last decade. He knew of our facility and the close proximity to Chicago and was hoping we would become one of his blenders and baggers. Not knowing all the aspects of the business, he asked if he could meet with the owners in the very near future? When I got back from Orlando, a meeting was scheduled to meet with the owners of Compost Product's and Joe. Before I left Orlando, I did tell Joe about my visit to Laurel Valley in 1973 and he assured me that they now had an open door policy.

All the owners and I met with Joe and his partner Peter Philippi from Germany in early March. All the advantages of Green Roof were explained, and Compost Product's became one of eight national blenders for the "Rooflite" brand of Green Roof material. Our area for production included all of Illinois, Iowa, Wisconsin, Missouri, Minnesota, Michigan, and Indiana. Chicago seemed to have the greatest potential because of Mayor Daley and his green thumb for the city. Part of the highly guarded formula contains a good deal of compost in the blend. We had the compost, we had the manpower, and we were very close to a lot of large metropolitan areas.

Once the material is blended, it is filled into super sacks that hold about two yards of material and weigh about three thousand pounds each. There are four eyelets on the bag so a crane can hook onto these bags and lift them up to the tops of the roof. Once on the roof, they are cut open, and the material is spread out to a depth of between four and twelve inches depending on what is to be grown. The main idea behind Green Roof is to absorb a two inch rain event without any water leaving the top of the building as runoff. It also has to be light enough not to compromise the structure with excessive weight. Water is held by the material blend for plants to feed off of until the next rain event. It will also keep the building warmer in the winter, and cooler in the summer, and the roof will last about three times longer than tar and pee gravel. Where land is very valuable, a lot less property is required for runoff detention because it is held by the Green Roof. It will be very good for Compost Product's because Chicago, Milwaukee, Minneapolis, Madison, and Indianapolis are the number one towns in the Midwest for Green Roof installation, and we are very close to these towns.

Very similar to the Gypost I was applying to the farm fields, I devised some very ingenious ideas on how to blend and bag this Green Roof material with little or no cost to Compost Product's. I got another taste of the new management after I never received any benefits from my innovativeness or my association with the old composter from Laurel Valley. In 2009, we were the largest blender and bagger in the Rooflite system. It was amazing to see some of the before and after shots of roofs we did in 2008 and 2009. Growing beautiful lawns and gardens on high-rise buildings in the middle of large cities.

More Machinery And More Memories

As far back as I can remember, I have worked with good people, and have supervised good people. I learned from the best while working for Gorman seventeen years. As large and diverse as his operation was, he sometimes would oversee as many as twenty men at one time. He would never have somebody do something that he wouldn't do himself. We all worked beside him day after day and as I stated earlier, Jack Tyler, a third generation grain elevator operator, made the comment that, "Gorman was the only farmer he knew that made money with hired help." I think I got a lot of my drive and innovativeness from him.

One very good example of innovation was harvesting popcorn for the Cracker Jack Company in the fifties, sixties, and seventies. Popcorn was traditionally harvested by taking the whole ear off the plant and shipping it to the popcorn factory. It was harvested with two and three row pull type pickers and eventually with self propelled pickers that would elevate the corn into wagons that were pulled directly behind the harvesters. The amount of work was horrendous and it all occurred out in the weather. The corn was harvested with the whole ear shipped to the factory because of the fear that damage would occur if it was harvested with regular, very aggressive combines of that time.

In 1969, John Deere introduced a self propelled combine where the threshing mechanism could be manually adjusted to very gently thresh the kernels off the cob without damage. Gorman ran this idea past the popcorn company and they agreed to give a couple of loads a try. We harvested and shelled the kernels off the cob in the field for the first time in November of 1969, sent it to Cracker Jack, and had amazing results from the experiment. It revolutionized the harvesting of popcorn and all that work out in the weather. I'm not saying we were the first in the country to do this, but this is how 95% of the popcorn is now harvested.

Another example of Gorman's ability to think occurred in the fall of 1979. It started to rain in early October of that year and never let up. We had over three thousand acres of field corn that needed to be harvested, but the ground was too soft and muddy to let the combines work. Gorman talked to a local John Deere dealer who said jokingly, "you need a set of tracks". These would replace the front tires on our combines and looked similar to what Caterpillars use. These were very common down in rice country, but had not heard of anyone using them in the upper Midwest.

The dealer knew of an operation in White County, Indiana that was owned by Danny Overmeyer, and was using tracks on their combines that Fall. Gorman asked me to drive the hundred miles and see if these devices were all they were stacked up to be. I had no problem finding this massive farm that encompassed more than fifteen thousand acres and employed over three hundred, mainly Spanish, men. Most of the help worked with the potato production side of the farm enterprise which needs a lot of help to produce and harvest chip type potatoes. The farm manager knew I was coming and met up with him at the farm office. The home base consisted of ten acres of machinery, twenty small homes for the year round help, and huge potato storage sheds. We drove about five miles to an area where corn combining was taking place. There was four John Deere combines all equipped with these track units going back and forth through this mud hole of a field having very little problem harvesting the standing corn. Some of it was in a foot of water and yet the combines were going right through this mess.

In addition to that spectacle, they had large four wheel drive tractors hooked up to large wagons that sat on two very large tires that could go out into the field, position themselves under the auger of the combine, and receive grain from the combine without ever stopping. These wagons were equipped with a large auger themselves so they could travel to solid ground and unload the grain into waiting trucks. These things were called "grain carts" and it was the first time I had seen them in action for myself. I told Gorman the next day what I had seen, and a truck driver was dispatched for the 650 mile trip to McCrory, Arkansas to pick up one of these track units for one of Gorman's combine. He also ordered a grain cart from the local John Deere dealer.

It took about three days to pickup and install the unit on the combine. The cost was $6,500 plus the trip down to Arkansas and back. Once it was installed, we took the combine out in the field next to the shop that was already harvested and found the worst wet spot. It ran through one side and out the other just like they were doing in Indiana. Most of the corn was located ten miles south of the home base in the Arsenal. So we hired a lowboy truck, hauled the combine down to the field, and started to harvest corn. About the time we started using the tracks, it stopped raining. The ground got a little firmer to the point that the second combine without tracks could even do some harvesting. In less than a week, we took the tracks off the combine because every time we moved to a different area we had to get that lowboy trailer to move the combine so it would not tear up the roadways. Gorman did call the Overmeyer operation and asked, "if they could use another set of tracks?" The next day I went back to Indiana with the track units and got a check for $6,000, about five hundred less than was paid. Gorman never sat around to see what might happened, he made things happen. Sometimes it cost him, most of the time it made him money. We did harvest all the corn that year when a lot of others were not so lucky. In the Spring of 1980, we were planting corn across from a field where the farmer was combining corn he could not harvest the year before and Gorman said, "you may never see sight again in your lifetime." So far, he has been right about that.

In the late eighties, I was back in the Overmeyer country only to find everything gone. They had ten acres of machinery back then, it was all gone. They had twenty or thirty small homes for the year round help, they were gone. They had three or four large warhouses full of potatoes to slice and fry into chips, not a single potato peel on the place. They had produced popcorn, spearmint, peppermint, alfalfa hay, and potatoes, nothing but corn now as far as you could see. A local farmer explained to me that everything went bankrupt in the early eighties. Land prices tumbled more than fifty percent, interest rates went up to twenty percent, and crop prices were very low. He even heard that a banker committed suicide over the huge amount of money that was lost over this one operation. I was never able to confirm that story, but I was told the same story by more than one farmer in that area.

Another big innovation in the late seventies was the invention and use of large corn planters. Gorman's whole crew took a day off in the fall of 1976 to attend the Farm Progress Show. It normally floated between the states of Illinois, Iowa, and that year near the town of Greenfield, Indiana, just east of Indianapolis. All the new developments and machinery ideas were on stage for all the farmers to see in this three day event. One very interesting display was a sixteen row corn planter engineered by a very energetic and industrious individual named L.Eugene Smith from Lebanon, Indiana. Gorman took one look at this and said, "of all people, he needed one of these planters." Most companies were producing six and eight row corn planters, and several were trying twelve rows that folded up in the air hydraulically to go down the road for transport. The limiting factor for large planters was the ease of transporting from field operations to road transport and then back to field operations again with very little hassle. Nothing was on the market like the machine Smith had developed by folding the planter backwards so it could go down the road at less than twelve feet wide and all done from the tractor seat.

Gorman placed an order for the planter frame right at the show. In February of 1977, Gorman and I took a trip to Lebanon to see how the frame was advancing and to pay half the price up front before delivery. We learned that Eugene had graduated from Purdue University in the Spring of 1968 with a degree in Agriculture Engineering. He received his Masters in 1969. In 1971, he bought a 100 acre farm very close to where he grew up with very little down payment, sold the buildings and a few acres for almost as much as he paid for the whole farm, and off he went. He bought and traded land in that area and even in Louisiana so by the time we met him in '77, he had accumulated over 18,000 acres of farmland, and was renting another 4,000.

His home base consisted of five or six huge machine sheds that housed all his equipment. He had built over two million bushels of grain storage on the farm. He had at least twelve full time employees that worked on all the machinery inventions Eugene came up with when they were not farming. Some of his help told us that Eugene can leave the farm for a day, and come back with ten new ideas for the help to chew on in their free time. His original 100 acre farm turned into 1500 acres in one single field by purchasing the land next to each other and clearing out the fence lines to form this huge chunk of land. He like to take perspective clients, whether it was for a land sale or a planter frame sale, around the outside of this farmland and show off this parcel of ground that laid just south of Lebanon right along Interstate 65. He took Gorman and I around, but we never got to talk to him much for all the incoming radio messages from his home base. That particular Spring he purchased over sixteen farms in the Lebanon area in less than a month. He used leveraging techniques learned in college and money borrowed from the Prudential Insurance Company to feed his addiction to purchase land.

The planter frame was about complete when we showed up on that cold February day. Eugene showed us another frame on the other side of this massive shop that would accomadate an unheard of twenty four rows. It was based on the same design as Gorman's, but another twenty feet wider. It was an amazing site to see and no camera around.

Gorman would now have to buy sixteen planter units off of John Deere to bolt onto this new folding frame. John Deere had gotten word about this simple farmer making large planter frames and told all the dealers not to sell units separate from the planter that particular year. Since Gorman wanted John Deere units, he was forced to go out and buy two new John Deere planters so we could rob the units for Smith's new planter frame.

The frame and two men arrived at Gorman's to help switch the new planter units off the two new planters purchased from John Deere and onto Smith's frame. Everything worked well. We opened the planter up and closed it several times, and Eugene's help left. We took the planter ten miles south to the Arsenal, unfolded it all from the tractor seat, and put some seed corn in the units. Gorman made the first pass down the field on April 20, 1977. It was quite a site. It was double the size of our old eight row planters, but it looked a lot bigger than that the first time down the field. When he got to the end, he raised the planter to turn around, and four hydraulic hoses blew with oil going everywhere. It was a Sunday morning, but we knew a shop that would open up and repair the hoses. I was back with the hoses in an hour and replaced the oil in the tractor, and off Gorman went with the second pass. He raised the planter on the other end, and blew four more hoses. I went through the same procedure as before, replaced the hoses and oil, and planted for about another hour before we had the same problem again with some of the other hoses.

Gorman called Eugene that evening and he indicated that if we were having that much problems that his help must have set the hose maker up so it put too much pressure on the fittings and the entire system might be suspect. There was over two hundred fittings on this frame and we could not count on any of them. Gorman decided to park the new frame, bring the two new John Deere planter frames down to the field, and switch the units back on to those planters. We finished the season with the two eight row planters and two tractors, something we were trying to get away from when we purchased the big sixteen row frame form Smith. In July of 1977, Gorman asked me to drive down to see Eugene and how we could rectify the problem with all those bad hose fittings.

When I arrived at Eugene's farm, we met in his office in his home. He had turned about half of the first floor into a business office with closed circuit TV to the Board of Trade, charts that showed corn and soybean prices for the last five years, and maps showing all the property he owned or was farming. He had at least three people working full time on paper work and other office tasks associated with farming on a big time scale. I asked if he remembered the frame problems we had in the Spring? He pulled out a large map showing me the seven thousand acres he had just purchased in Louisiana. He would need a lot of equipment to run this land and asked, "if he could by Gorman's planter frame back for the same price as he paid the year earlier?" I told Eugene, "that was an excellent idea." I found out from one of his employees a year later, that all hoses and ends were replaced, and they used it very successfully on his new land purchase.

While I was on the farm, Eugene wanted me to see something in the big shop. He was fabricating a planter that would plant thirty six rows at a time and planned on using it himself the next year for all of his ever increasing acreage. I don't believe Eugene made any more than about twelve frames in his career mainly because of another planter frame idea that was brewing out there and huge financial problems that were coming very soon.

I followed Eugene's career for more than twenty five years. His pace of expansion was chronicled in many farm magazines. When there was an article about him, I would cut it out and started a large folder all about him. He was the undisputed king of expansion. It all came crashing down a lot faster than it went up starting in 1980. Interest rates went up to twenty percent and land values were cut in half. Before that, land appreciated so fast that he bought land with all the paper appreciation from his other properties and the banks were loaning money with nothing down. Boy that sure sounds like what happened to the housing market in 2007 and 2008 in the U.S. Eugene had in excess of twenty thousand acres and lost everything in two years. He told one reporter from the Wall Street Journal that he lost $1,000 per hour, twenty four hours per day, everyday for two years. Over fifteen million of equity lost. He officially went bankrupt the last week of December of 1982.

The last nine thousand acres, which included his home base, all the grain and machinery storage, was deeded back to Prudential Insurance for the twenty million in return for what he owed them in 1982. They in turn, rented it back to him in 1983 for ninety dollars per acre. The government, because of the PIK program, paid him over one hundred and thirty dollars per acre not to farm the ground that year! He pocketed over a million dollars with out planting a single kernel of corn that year. He was the only farmer in the Midwest to collect that much money from the government without farming a single acre. He also had about one million bushels of corn in storage that the bankruptcy courts did not want because the price of corn was so low in December of 1982. When the government announced the PIK program, corn jumped more than a dollar and Eugene sold all that stored corn for about one million in profit for future delivery. He started to speculate on the board of trade, ran his two million up to almost five million, was thinking about buying his old home base back, but lost all of the money when he guesed wrong with the soybean market over the next six months. .

Eugene tried to run for political office as a representative from the Lebanon area to Washington in 1986, but blamed Prudential Insurance Company for his loss in the election because he had to move a lot of that corn from the 1982 debacle off the farm and could not concentrate on the election. He was going to go to Washington to try and change what he felt the government and Paul Volcker did to destroy his empire that he had built in those few short years. He felt Volcker turned him upside down and shook everything out of his pockets. In 2001, he died on the historic Ribeyre Island Farm near New Harmony, Indiana working on an electrical problem with a center pivot irrigation unit. Most people that knew Eugene either liked him or hated him. I have to put him on my list of most respected and it was certainly a great pleasure knowing him.

The very next year, after Smith bought his planter frame back from Gorman, another innovator named Gene Shoup had designed and built a sixteen row planter frame about twenty miles south of Gorman's home base near Bonfield, Illinois. His ingenious idea was not to fold the planter backwards like Smith did, but he folded the planter frame forward towards the tractor for easy road transport. Gorman was the first farmer to contact Shoup even before he did any advertisement. An order was placed and the frame was delivered well before planting time in the Spring of 1978. We had to take all those planting units off the two John Deere planters and put them on the new Shoup frame, and the rest is history. We never encountered a single problem with this planter, and used it until the two ladies sold it at Gorman's auction in 1984.

Gene Shoup told us the story of how John Deere stole the idea of front folding corn planters from him in 1977. He contends that a black helicopter followed him for many days during the corn planting season showing up for an hour then disappearing for a couple of hours and then reappearing about the time he finished a field and was ready to transport to the next field. When John Deere first introduced their version of the front folding planters, Shoup decided to sue them for patent infringement. Gene never told us about the verdict, but we heard he did very well and went on to produce several hundred of these frames for farmers all across the Midwest.

I did not know until after Gorman died that Gene never took a dime from him for the first commercially built front fold planter frame. He considered it free advertisement that a farmer like Gorman would use one of his frames and hoped by word of mouth it would quickly spread through out the farming community. In the Fall of 1981, we custom harvested about 580 acres of corn for Gene near the town of Elwood, Illinois. It was after Thanksgiving and was getting late to be harvesting. Gene called Gorman to find out if he could help with the harvest? Despite having all the harvesting equipment cleaned up and put away for the year, we were in the corn field the very next day with all our machinery and all our manpower. When Gorman died six months later, I learned from the two O'Reilly ladies he had never gotten paid from Gene for the custon harvest of all those acres six months earlier. I never found out if Gene called when he heard Gorman died or if Gorman had it written down somewhere. Anyway, I think it was Gorman's small payback for not charging for the first front fold corn planter frame four years earlier. Gene went on to become very successful not only with the planter frames, but also his own company known as Shoup Manufacturing selling replacement parts for all the different manufactures at very reduced prices.

Machinery has always been a huge part of my life whether it was for farming or composting. I grew up surrounded by all John Deere equipment which made it very easy to get parts by going to just one place. The machinery for composting were all special built and required a lot of different vendors to find replacement parts when something went wrong. I hate to harp about it, but the old mushroom composter never gets enough credit for all the equipment we take for granted today.

At the turn of the twentith century, most compost got aerated by using a fork and a strong back. With the advent of engines and hydraulics, ideas about machines started to evolve. The first machines for turning was the choke feed that I described earlier. The first self propelled turners blossomed in the early fifties. These rectangular shaped machines were about ten feet wide, about ten feet tall, and about thirty five feet long all made of steel. A large engine was mounted on the top of the platform and had numerous pumps attached to it to run motors for propulsion and rotating large drums. The large drum in front would rotate upwards at about forty rpm's picking the material up off the concrete and and delivering to a second smaller drum behind that rotated at about four hundred rpm's. This second drum would vigorously throw the material rearward to reform the straight sided ricks while the machine was traveling forward. The job of providing oxygen to the compost had been completed with the use of a self propelled compost turner and much easier and faster than the old choke feed machines.

On my first trip to the mushroom capital of the world I met Bob Pannell of Kennett Square, Pennsylania. During the mid seventies, Bob was one of the first to come up with an idea to reduce the amount of anaerobic material in the rick of mushroom compost. Despite the amount of turning, a large area called the "core", would develop in the middle of a lot of compost ricks that stayed anaerobic the entire two week process. It would stink and mushrooms did not grow as well on that type of compost. The early self propelled turners would pick the material up off the concrete pad, and deliver it right straight through the turner to reform the straight sided rick behind. The outside would stay on the outside, and the inside would stay on the inside of these ricks. Bob's patent was to take the single drum that turned at four hundred rpm's and bolt directional paddles on it so when the material from the front drum was deposited, the outside of the rick would be thrown to the inside, and the inside of the rick, would be thrown to the outside. The first single drum "cross mix" turner was developed and a lot less stink and a lot better compost was made because of this idea. He took the machine one step further by offering an all stainless version which were just about indestructible. The only good thing that came out of the nine month relationship with Money's mushroom was the purchase of one of those stainless cross mix turner from Bob. I thought I had reached the pinnacle of composting when we received one of these rigs in the Spring of 2000.

Unfortunately, we were only able to use it to make compost for Prince Crossing for about five months before Money's declared bankruptcy. Bob and his family can be added to my growing list of most respected in my forty years of composting. He and his son, Bob Jr., still laugh when ever I call them remembering the story from the first time I made a visit in 1973. I tried to get into a compost site, it was one of Pannell's bigger customers, to ask some questions and see how compost was made. I was thrown off the farm without a single question being asked. There was a road around this place that I discovered you could drive to and from a high spot watch what they were doing in the compost preparation area. After viewing the operation for about ten minutes, the farm called the local police and was asked to leave the area.

My next stop that day was Pannell Manufacturing about a mile down the road. I told them the story and after they stopped laughing for about ten minutes, they showed me around their shop and everything they produced for the mushroom industry. After the tour, Bob said he wanted to show me something. We loaded up into his truck and drove right back to the site where I had just got thrown off. He drove right past the main entrance straight back to the compost prep site so I could watch and learn about using all this Pannell equipment. Seems like Bob had full access to any site in the area and the men who threw me off the farm about an hour earlier never knew I was there. It has always been big joke for Bob and his son.

Before I go any further I have to describe the worst machine I was ever near in my forty years of composting and farming experiences. While we worked for Campbell's, we would take raw feedstocks used in compost preparation and gradually bring the moisture up to about forty percent. At that point, we would run some of the runoff water from the retention pond to a low point in the concrete and dip the material with the big pay loaders trying to get the material to seventy five percent moisture. We would then apply about half of the nitrogen source to the material, pick it back up with the loaders, and make a big pile to start the compost process. After a couple of days, the material would then be loaded into the choke feed turner to form the initial rick for the self propelled turners to handle. This began the fourteen day cycle of mushroom compost.

Money's mushroom was convinced that we needed a machine to do this job instead of using the loaders and all that free water. It was termed a "pre-wet" machine and everyone in the industry that had one of these very large, cumbersome, and very expensive machines wished they had never bought one. It was made in Canada, it sat on two large very wide wheels in the front, it had two small caster wheels on the back for turning around. A large rotating drum about fifteen feet wide would pick up the material off the concrete and bring it to the center of the machine so it could be watered and elevated up and back to form a pile of material behind the machine. A large engine sat very low under the elevator. When it was operating that caused a lot of fires we had to watch for constantly. It created double the work for the loaders, water ran off the material carrying a lot of the nitrogen supplements with it, and it increased the time to pre-wet a house of compost by two times of what it normally took. I tried to talk to Money's mangement out of one of these machines before they transferred it here from the now closed Dublin, Georgia farm, but was unsuccessful.

I don't believe I have ever been around a more poorly designed or engineered machine in my entire life. I grew up around John Deere all my farming days. I had met Bob Pannell and used his very well designed and built turning equipment for the mushroom industry. I had met and worked with L. Eugene Smith and Gene Shoup, great engineers in their own right. I had never seen anything like this. The owners of our operation told me after using it for about two weeks, to park the thing in the corner and lose the key's. It was shortly after I did that, Money's declared bankruptcy. Not a single pound of all that special pre-wetted material ever grew a pound of mushrooms before they closed the farm down for good.

When Money's came for all their mushroom equipment in the Spring of 2001, a crew of roughnecks, a crane, and a fleet of lowboys showed up unannounced to load all this machinery and move it off the property. There was so much damage done to every piece of equipment from chains, torches, hand grinders, and improper loading techniques, that I don't believe anything was ever put back together and used to turn mushroom compost at any facility in North America. Even my prized stainless steel, cross mix turner made by Pannell was the victim of this unbelievable crew of roughnecks. It officially ended our mushroom compost era.

When I made the trip to Holland, Michigan to see the Vermeer grinder we bought, I stopped by the old Glenn farm for a few memories. The only thing left on that huge concrete compost pad where all that compost had been made for all those years was one of those pre-wet machines standing there like a statue. It seems nobody else wanted the thing at any price and is probably being used as a cattle rub by the dairy that bought the property.

More People And More Problems

During my forty years of composting and farming, a day never went by that I didn't learn something. It pertained to the people I worked with, the machinery I used, the process of making compost or growing crops, and interacting with customers. In dealing with people, I have worked with self starters and non-starters, the hard worker and the lazy worker, the honest person and the liars, on time help and always late help. I have had workers you can tell something one time and it is done to perfection, and help that it seems to be their first day everyday.

After Gorman died in 1982, then and only then did I realize the magintude of his management capabilities. He was there day after day to lend a hand with the overall operation we had going in Joliet. When I took over the entire responsibility for Compost Product's, I adjusted very quickly in order to keep Campbell's happy. I rose to the challenge without a problem and dispelled the notion that everybody rises to their level of incompetence.

From 1983 to about 1990, Ramon Barajas was the growing manager at Prince Crossing where we sent all that compost for so many years. Mushrooms require a lot of tender loving care after the house is filled with a new batch of compost. Ramon supplied this TLC everyday to each and everyone of the sixty houses that made up Prince Crossing Farm. Everyone said that Ramon had gone through every house by seven in the morning everyday, even on days he was off. He would make notes about certain things and relay them to his underlings about what necessary adjustments should be made. Maybe a house needed to be cooler or warmer, maybe more fresh air, more carbon dioxide, more water, a deeper casing layer, a longer spawn run, estimated time to pick the days harvest, and about one hundred other things about all the houses. Nothing was missed by Ramon those first few hours of everyday and even came to see us, over forty miles away, at least once per week.

The farm was doing great and Ramon was promoted to be the farm manager of Prince Crossing. He was the next in line and he certainly deserved it. His responsibility was expanded to include all the phases of the normal mushroom farm; filling, phase two, spawning, spawn run, casing, harvesting, packaging, office personnel, vendors, and customers. He could not just concentrate on growing anymore because there are not enough hours in the day. Production started to decline from the normal 250,000 pounds per week, to less than 190,000 pounds per week. Some of the quality was also lost during this three month decline.

At first, all the attention was directed to the compost production phase, what we must be doing wrong. The old rule was if they get a good crop it's what the grower did, if they get a bad crop, it's what the composter did. We started making a lot of changes as far as amounts of supplement, when we put them on, longer composting time, more turns, temperatures were watched more closely. A lot of changes, and in my estimation, none of it was needed. We got so sick and tired of all the new changes ordered, that I approached the owners with an idea. If Campbell's would go along, I wanted to prepare a house of compost and ship it to another farm in the Midwest and see how they would do with our compost compared to what Prince Crossing was doing. In return, that farm would send one of their houses of compost to West Chicago to test what Prince Crossing could do with their mterial.

We got the go ahead and we made a house just like always, and sent six trucks loaded to Glenn, Michigan farm. Once they got dumped, they were reloaded with a house of compost they had produced and brought it back to Prince Crossing for the test to begin. Since the mushroom cycle takes about ninety days, it would be a slow but very decisive test to get to the bottom of the problem and to prove there was nothing wrong with what we were producing at Compost Product's. We were putting our money where on mouth was so to speak. Before the experiment was completed, the upper management in Campbell's decided to lessen Ramon's duties as the farm manager and made him assistant farm manager so he could spend more time back in the mushroom houses. In less than a month, production climbed above 200,000 pounds.

The lesson learned that came out of all these changes was that people will rise to their level of incompetence in any arena. We had very dedicated underlings working at the mushroom farm that were put into the growing position just a little too soon and production suffered. The experiment did conclude, and the compost we sent to Glenn did in fact produce more mushrooms than their own indoor compost. We had a big infusion of compost personal and management from all the other seven farms visit our facility over the next two months to see what we were doing different to have good production. It felt real good and it certainly helped with my contract extension with Compost Product's the next Spring. Ramon Barajas can also be added to my list of most respected.

Through out the nineties, Campbell's called on me to do some trouble shooting at their compost operations around the country. Problems usually came from some very simple mistake that was being made in the compost process. The compost would end up being too long or too short, it was too wet or too dry, or it was very uneven as far as the maturity. A lot of people would look at the finished product going into the mushroom house and try and determine the problem. I would always start at the beginning where the raw ingredients first came into a farm. My idea was if you put garbage in, you will get garbage out.

Starting with the feedstock's, the stabel bedding, hay, straw, cobs, cottonseed meal, chicken manure, gypsum, urea, poash, and water, I could very quickly determine if there was in fact an inbound problem. Sometimes you could end the visit right there at that point. A lot of times a crew working on the same site for years might not see a problem that a greenhorn to the farm might see and question in a minute.

If everything looked fine to this point, I would move on to the area where everything got blended together for the first time. You have to mix ingredients that are about the same age of decomposition to get an even finished product. Then to the area of building the initial rick for composting. This is where the rectangular shaped loaf like pile does most of the composting. I would observe how the material heated up, is the material even from one end to the other, make sure the material was at the proper moisture, was the supplements evenly distributed and about a hundred other observations all connected to good compost production.

I also observed how the machinery was working, especially watching for oil leaks. Hydraulic oil and mushrooms don't mix and hydraulic turning equipment that has a lot of oil leaking will effect the crop. To this point, if nothing looked wrong, I tested another secret from when I first visited the Pia operation in Kaolin, Pennsylvania back in 1973. Take a large handful of compost from the oldest compost, and squeeze it between your hands until, hopefully, you get water to drip out between your fingers. Depending on how hard you have to squeeze, you will determine the moisture without having to test for it, and you can reveal something else about the entire operation. When you wash your hands off and about two hours later you can still smell the odor of compost on your hands, you have discovered that the compost is being made under too much anaerobic conditions. Mushrooms and neighbors don't like compost made under anaerobic conditions. There are two ways to reduce the amount of anaerobic material. One is to make the ricks a lot narrower for better oxygen flow, or second, you can use the cross mix turner developed by Pannell to get the inside of the rick to the outside, and outside of the rick to the inside. I have seen remarkable results when the farm has discovered the amount of anaerobic compost that was being produce. A lot of the operators will not notice because they do the job everyday the same way and could not see it like a newcomer to the site could.

Many times a farm manager would fly me in to his farm and pay me for my time knowing there was probably nothing wrong with the composting process on his farm. He was just looking for some insurance and to bolster his compost crew that it was not their problem of low production at the farm. There was always a lot of apprehension associated with somebody foreign coming into your farm to critic your job or operation. I tried to be as professional as possible with the consultants job and really got along well with all the farms and their staff. The world of composting is still a very small family. The number of composting sites in the U.S. is less than 2500, and in the world, less than 5,000. The use of a consultant from time to time is very important and very cheap tool to use.

After Compost Product's made the decision to make composted landscape mulch in 2000, I set out to try and find out how to do this successfully. It brought back a lot of memories of when I did the same thing in 1973 to make mushroom compost. When I started to look around in 2000, I was not very well received because I would now be competing for the same market these other operations were selling into around Chicago. I attended a compost conference in Columbus, Ohio in November of 2000 shortly after Money's went bankrupt and left us high and dry. We were still wondering whether we could make a product like spent mushroom compost and sell the product.

I had the great fortune to meet Sharon Barnes from Huron, Ohio. She and her family were the trend setters for windrow composting and all the uses for compost in the state of Ohio. I received a lot of confidence from talking with her that we could do the job. We had a great area to assemble feedstock for processing, and we had the best market in the Midwest to sell finished material. She and her family are tremendous role models for her state and the rest of the U.S. as far as composting and innovation. There is no doubt that Sharon and her leadership is one of the main reasons the U.S. Compost Council is what it is today and so successful. They are certainly someone I can add to my list of most respected. Everytime I see her at Compost functions, I never fail to thank her for all she did to help us move seamlessly from mushroom composting in ricks to windrow composting for the landscape market.

Our initial formula for windrow composting started off with 100% stable bedding from all the Chicago racetracks. What ever could not be sold to Monterey Mushroom in Princeton, Illinois, would be dumped on me to handle and compost. The history of ownership had changed again at the big mushroom farm with Terry's Mushroom going into bankruptcy and Monterey purchasing the farm in and around 1998.

The volume of stable bedding has always been determined by the weather in Chicago. The colder it got, the less horses, and the less stable bedding. The volume from the five racetracks could vary from three hundred yards per day at the lightest to over fifteen hundred yards per day at the heaviest. There was also the large horse farms that needed to be cleaned out and would add another three to four hundred yards per week. During the hay day of racing in Chicago, there were over six thousand horses boarded in the five racetracks and the large horse farms at one time.

A small portion of the horses were boarded on woodchips or sawdust instead of wheat straw. The owners of our farm asked me if we could use this material? So I did a little experimenting. The finished product looked great even with the addition of up to fifty percent wood based manure. We were getting paid too much to turn it down and the K/W and the windrow system made a beautiful product out of this new feedstock. We also took in a lot of leaves and again made some very nice compost out of those. I had forgot about the importance of the C/N ratio until one of the neighbors asked me if we had shut down because there was not the usual odor as when we made mushroom compost? We were probably running at a 50/1 C/N ratio which meant very little odor but a lot slower composting.

Speaking of odors, we did have one large odor problem with making mushroom compost back in 1997. Despite the best equipment made by man, mushroom production does stink. When anybody wanted directions to our farm I would always tell them, you will smell us before you see us! Odors were the main reason Campbell's could not compost at their backdoor getting pushed out of West Chcago in 1975. Mushroom compost starts out at about 14/1 C/N, and only gets worse in the two week process. You add to that fact that there are fifteen or so ricks going at one time, and there is a lot of smell produced. If the wind blows the wrong way for a very long time somebody could complain, but it was not any worse than a hog or cattle farm in the area in our estimation.

We had been making compost at the LaSalle County location for about fourteen years when a neighbor, some three quarters of a mile away, sued us for odor nuisance. They campaigned in the area and got over thirty five signatures to have us closed down. Some of the names on the lawsuit were from over three miles from the compost site. The farm sat just inside LaSalle County, but most of the neighbors resided in Kendall County and that is were they filed the motion. Our lawyers spent the first two years trying to get the motion moved to either LaSalle County where the farm was located, or DuPage County where our business headquarters was located. After we failed to get it moved, a two day trial did take place in 1999 in Yorkville, Illinois, the Kendall County seat.

The plaintiff's attorney was a very close friend to all the people that signed the original petition to have us closed. He never bothered to depose a single person on our side figuring it was a slam dunk to have us closed. I think he also knew the judge real well since he was before him a lot in that county. The number of people on the complaint dropped to eight after the lawyer needed some money to continue the case.

We supplied expert testimony that in any given year, any of the people who were complaining could only smell the farm less than one percent of the time, and on any given day, for less than a twelve hour duration. We had testimony that we were in a livestock producing area and smells are prevalent all the time. We had expert testimony that no person was ever sick from the odor. We had tested the wells from over ten residents when we built Compost Product's in 1983 and tested them again in 1999 to reveal there had been no change to their status. Campbell management testified that the jobs of over 250 people relied on this operation for their lively hood. Since I was not allowed in the courtroom because I was going to testify and be asked questions, I was told by the owners that the judge seemed to be sleeping the whole time our witnesses were up on the stand testifying.

I was called to the stand the second day in reverse order of what our lawyer said was going to happen. The plaintiff's lawyer called me and treated me as a hostile witness asking a wide range of questions. He had not depose me before the trial started and anything I said could come as a surprise. He asked about my experience with compost, and I told him I had been making it for almost twenty-seven year's. He asked a little about the process, and I explained exactly what we did and I said it was an operation second to none. He grilled me immediately on how I would know such a thing? I told him, and it seemed at that particular moment the judge woke up, that I had been on over one hundred compost sites in the U.S, Canada, and Mexico in the last twenty-seven years. Our farm was the best facility I had seen, we had the best equipment, and we never varied from our schedule, even working seven days per week. In addition to that, we had received the Select Supplier Award from Campbell Soup years earlier for our process.

After he was done putting his foot in his mouth, our lawyer only asked a couple of questions about livestock operations in the area, and whether I thought we were doing the best job possible now since there had been no complaint for the first fourteen years? A few of the plaintiffs were called to the stand with one hog farmer we knew stating that hog manure does not stink. Even the judge laughed at that one. After the two day trial, the judge indicated he would have his decision in the next thirty days. On the thirtieth day, the judge found in favor of the plaintiffs, and we were to be shut down.

None of us could believe the verdict. We felt they did not prove their case, no one was ever sick, the exposure was minimal at the worst, and we had been there for over fourteen years. Our lawyers did ask the judge to let us continue to operate while we appealed the decision, and that was granted. There are very few appellant lawyers in the crowd, but luckily one of our lawyers of the three we had was and immediately went to work on an appeal in the first district in Wheaton, Illinois. That took over a year before the case was heard without any testimony given and nothing new added. It was about this time that Money's went bankrupt and we would no longer be producing mushroom compost and a lot of the smell the plaintiffs were complaining about. The Appellant judges decided in our favor and set aside the verdit of the lower courts. In addition to that, they imposed sanctions on the judge for presiding over the trial unfairly. I understand that he was dismissed from his position shortly after our verdict was overturned.

The plaintiffs had thirty days to appeal to the Illinois Supreme Court, but decided to refile the same motion in DuPage County, exactly where we wanted the trial to be held five years earlier. Both sides agreed to have no testimony given using all the court papers from the previous trial from Kendall County. The judge decided in our favor. The plaintiffs had thirty days to appeal that decision and they did, going back to the same set of Appellant judges that heard our appeal about two years earlier. The Appellant judges ruled in our favor again and the plaintiffs did not pursue it any further at the Supreme Court level. Late in 2003, we were free from a seven year odor nuisance case and about $325,000 in court fees, expert witness fees, and lawyer fees.

After being around compost or farming for over forty years, smells or odors are a very natural part of the business. We would do a lot of school bus tours when I farmed with Gorman. The kids would come out to see the livestock and all the machinery in the late Spring or early Fall. Everytime the kids got close to a large beef cow, it would naturally need to relieve itself of pee or poop. You could almost bet on it. It was always very funny to hear them groan from the site and smell, but it was the natural way of farm life. We did the same for the kids when we started Compost Product's. They would enter the site and get off the bus with their hands over their noses. It would only be a few minutes and most of them could not wait to get their hands into the compost and never complained about the stink after that. It was just a matter of time, and would happen every time without failure. It's what you get used to.

I did have a near death experience with a very unnatural smell of big time farming in the Spring 1976. Nitrogen is needed to grow corn successfully. Anhydrous form of nitrogen is the most widely used, but also the most dangerous. I was applying anhydrous ammonia to some land in the Arsenal for corn production that year. The job requires a large tractor pulling an implement that has twelve or more large knives that deposit the nitrogen in the ground to a depth of about twelve inches. There is a two thousand gallon tank pulled behind the implement with the tractor that supplies the ammonia to the implement and into the ground. When ever any ammonia escapes, a little white cloud will form, but for proper application, one should never see this. I was applying the nitrogen late in the afternoon and the large hose that runs between the tank and the implement ruptured sending out a cloud of ammonia about the size of a two story house. The cloud quickly engulfed the tractor to the point I could no longer breathe. I stopped the tractor and bailed out of the cab trying to run upwind from this growing cloud of ammonia. The tank was close to being empty but took nearly five minutes for the broken hose to let it all out of the tank.

Gorman was planting about two miles away, but could see this unusual white cloud in my general area. I could not get to the FM radio to call for help, but in about twenty minutes Gorman showed up to see what the problem was. I told him that I was extremely sick, and needed to get to the hospital. He shut the tractor off, helped me get into the pickup, and drove me the twenty miles to the hospital. I was administered oxygen and IV fluid for the next thirty six hours. The worst part was I lost over thirty four pounds sitting on the toilet the day and a half I was there. Since that day, I have been deathly afraid of electric, under ground pipes, and now anhydrous ammonia.

Stable Bedding And Customers

The life blood of all mushroom production has always been cheap and readily available feedstock year round. The problem with growing mushrooms near Chicago after 1975 was the competition for satable bedding because of the big farm Ralston Purina built near Princeton, Illinois. Campbell and the small family operations had free rein on the supply up until that time. Our sister company, Compost Supply, took over the removal and hauling of all the racetracks after Ralston fell on their face in 1978 and Fred Noorlag purchased all the equipment for twenty cents on the dollar.

Stable bedding volumes for everybody were fine for nine months out of the year. When December and cold weather hit, it became very ugly. Our company shipped to the Princeton farm first, year round, what ever they wanted, because they were willing to pay top dollar and sign a contract. We were also in a good position to remove some of the spent mushroom compost from that farm while we there. Campbell soup, being on the short end, put out the word in Lexington and Louisville, Kentucky that they needed to maintain a three week supply at Compost Product's, and they were willing to pay good money for material delivered to our farm during the Winter months.

Every type of truck and trailer started to arrive and at every hour of the day to unload in the Winter months for the first few years we composted in LaSalle County. The load looked like a certain yardage when it left from down there in Kentucky, but after the 400 mile journey, most loads settled down and shrank by fifteen or twenty percent. Fred even sent his own trucks down to Lexington, but it got rather expensive to do so. Early in 1987, Fred got together with Tom Creech from Lexington, Kentucky to come up with a better solution. Tom was sourcing a lot of this loose material from all the stud farms in that area that showed up at our farm.

Lexington was the horse breeding capital of the world and from December to April there was plenty of stable bedding, or as they call it, "Muck", that needed a home. We were short of muck in the winter, he had plenty. The only problem was the four hundred miles that separated us. Tom and Fred put their two ingenious minds together and decided to use a garbage baler set up somewhere in the Lexington area to bale muck and ship it north to the mushroom farms. These four foot by four foot by six foot long bales would be compressed with over a hundred thousand pounds pressure and tied off with five number nine wires. Each would weigh at least one ton and could be loaded on any truck that needed a load back into the Chicago area.

To see if their idea would work, they loaded and shipped two loose loads of muck from Lexington, Kentucky down to near Atlanta, Georgia for them to see a baler in operation bale this material before the actual purchase. They made twenty eight bales from the two loads and shipped them back on one refrigerated van that was headed to Gary, Indiana. Logistically, they were already a winner taking two trucks and turning it into one truck load. When the load arrived in LaSalle County, it was about ten degrees outside. When it was baled in Atlanta, it was sixty degrees. When we opened the door to remove the bales it was like a steam bath, not able to hardle see your hand in front of your face.

We did not have a fork truck to off load the one ton bales, so we hooked together a lot of chains, and pulled each bale out of the trailer individually. We set them up in the corner of the concrete and opened up two bales everyday to find out if this heavy compression did anything to hurt the material for future composting. Campbell's thought that being baled so tight without oxygen, that the material might not act like regular stable bedding and compost properly. All that worry was quickly eliminated when we opened the bales up. They took right off like they had never been in any form but loose delivered material, and we now had material for the Winter at a very reasonable cost.

After the first bales were produced, some of the other Campbell farms jumped on board and started to order baled muck from Lexington. Creech went ahead and built a first class baling facility by himself after we found out Fred was so sick and died in Fall of 1987. He continues that operation today sometimes producing more than two hundred bales per day for delivery to Pennsylvania and Florida. Become a millionaire from baled muck, who would ever think of that?

Creech, as most people refer to him instead of Tom, started out as a kid delivering more than four hundred newspapers before school each morning. When he got old enough to drive, he bought a small dump truck and started to deliver coal to residents and businesses in the Lexington area. He eventually moved into the horse service industry by purchasing hay and straw off of reliable producers around the country and shipping it to the horse breeding capital of the world and selling it to the stud farms in Lexington. His cliental grew very rapidly because he delivered quality product, it was there when he said, and his checks were always good. He is one of the biggest hay and straw buyers and sellers in the U.S. and a major exporter to England and the Midle East.

I got to know Creech a lot better after Fred died in 1987. He was forever bouncing ideas off me since I was associated with a lot of the people in the mushroom industry. He was not only selling hay and straw to the stud farms on the front end, but he was cleaning up the muck from the farm on the back end. His customers for the baled muck were also increasing through out the mushroom industry. The biggest advantage with the bales were that the volume was the same for every single bale despite the weight. If you needed six hundred yards, you could divide by six and a half yards per bale, cut that number of bales to make up your formula. That has never varied.

Creech had a lot of ideas for muck. Come to think of it, he has a lot of ideas about a lot of things. When he calls, you better have at least an hour of free time to talk about something he is thinking about. With muck removal, some of the very driest and cleanest is delivered to a local farm field where it is spread out into small windrows and rebaled with a small farm baler to use along roadsides to establish new grass seeding. He invented a large truck mounted spreader that could take the one ton bales of muck he makes at his bale plant and spread them on strip mine land for reclamation and grass establishment. He has also been very ingenious with the sale of hay and straw to the local stud farms and how it is delivered and handled.

In 2001, things changed for Creech. Our old farm in West Chicago was shut down. The farm in Jachson, Ohio was shut down, both by Money's. Our company could haul excess stable bedding out of Chicago over to Glenn, Michigan and Brighton, Indiana cheaper than Creech could bale and haul his muck to either farm. So a lot of muck was being produced in Lexington with no home a good portion of the year. He started to charge more for the removal from the stud farms, but the volume did not slow down at all. He invited me down to see if windrow composting for the landscape market would be an option for his company.

I spent three days in the Lexington area to canvas for a site, and what the climate was for marketing this material. The soil in that area is made up of a lot of brown clay loams and looks like it could use a lot of compost. He had picked an old site near a landfill that had been closed for about two years, but was not capped yet. The city of Lexington owned the property and was willing to lease some of the land for a compost site. Our motive with the city was we were cleaning up all the waste being produced at all the farms in an environmentally friendly way since about fifty percent of the market for muck had disappeared in the last year and something had to be done with it.

There were no residents within about a mile of this site which would really help with any odor complaints. His dream was to compost all the excess muck and sell it back to the city for capping the old landfill right next door. While I was there, we approached the city with the proposal to take all the bio-solids produced by Lexington, mix that with the muck, compost the two together, and use that as capping and closure material. We figured we could save the city over two million per year even after paying Creech a tip fee, and could stop dumping the bio-solids in a landfill some forty miles away.

We drove out to the site where Creech wanted to put this compost site, and immediately got stuck. I would have never dreamed he could put up such a fabulous place in this God forsaken country. He leased thirty acres off the city and put down ten acres of concrete the first year. Just about double from what I said he would need. He purchased a windrow turner twice the size and horse power he needed. As a matter of fact, everything he did was twice as big as I had suggested. It was a little like Fred when we started building Compost Product's back in 1983. Creech figured it worked for Fred, so it will probably work for him as well.

Creech never got to cap the old landfill because they reopened with a permit to increase the height. The city never took him up on the bio-solid removal for composting and still haul it forty miles to a landfill. Creech has always wanted me to come down and run his compost operation for him, but have never pulled the trigger to do so. Tom Creech and his entire staff can be added to my list of most respected.

The life blood of mushroom compost or any compost is feedstock's. The life blood for compost is a customer. For years are only customer was Campbell Soup. Their name changed about three times while we worked for them before Money's took over. A lot of changes, but the man in charge of the division through all those name changes stayed the same for a long time.

Dr. Bob Miller headed up the mushroom division for Campbell's, and I think there was only one person between him and the CEO. He was an Iowa farm boy that made it all the way to the top. He crowned "me compost when compost wasn't cool." He was very respected in the Campbell division as well as the entire mushroom industry. When Miller spoke, people listened. He had ideas that were always way ahead of their time. Campbell's tried to sell hydroponic lettuce, tomatoes, cucumber, and bell pepper along with mushrooms for a while. He thought the lettuce should be chopped up, mixed with some carrot and red cabbage, and packaged in one pound packages ten years before it became very popular. He wanted the division to concentrate a portion of their houses on growing ugly mushrooms like Portabella, Shitake, and Oyster and sell them for big money instead of that beautiful white button mushroom that everybody was growing and selling for next to nothing. He said that someday, "one of the biggest problems in the U.S. will be a shortage of truck drivers." With all the testing and drug screening he was sure right about that. He always got on me as a farmer telling me the biggest problem with farming was not adding value to your crop being produced. That was well before we started using corn for ethanol and soybeans for bio-diesel. Those two things sure added value to the crops in a hurry.

I always like Dr. Miller and enjoyed being around him. I would probably see him two or three times per year, usually at one of the conferences that Campbell's were always having. The growers and the composters would get together at a fancy hotel near one of the mushroom farms to discuss problems and solutions. I was always invited because we supplied Prince Crossing with their compost. I will never forget the meeting we had near the Dublin, Georgia farm.

It took place in a county that was dry and most of the men at these meetings never shyed away from a drink after a hard day at the mushroom farm. Dr. Miller gave me three hundred dollars and asked me to drive across the county line and fill the trunk up with some liquor for the men that night. I had to go to the town of Douglas and got a little nervous about bringing this much liquor back to another county. I thought about the movie, "Macon County Line" and felt a lot better after I got back to the hotel and got all the liquor up to the room. We had the hotel staff fill up the bathtub with ice in a hospitality room and I filled it up with beer and some hard liquor to be enjoyed the next two evenings. We had a lot of good times, but a lot was accomplished in the two or three days we got together.

When Dr. Miller announced his retirement in 1999, his replacement did a lot of damage to the division in less than a year and ruined a lot of Dr. Millers' fine work he had accomplished over his long tenure with Campbell Soup. It was not long after that, Money's came in, purchased the mushroom division, and went bankrupt. I hear that Bob lives in Florida and plays golf everyday of the year and no one is more deserving. I can certainly add him to my list of most respected.

After the days of mushroom compost were over, we had an explosion of customers for our composted landscape mulch. We had customers that came with a five gallon pail, and we had customers that took a hundred semi-loads each year. We also started to spread compost for the farmers as a customer, but that only lasted one year. I never thought I would ever see that even if it was for just one year. We had customers for the green roof material that went all over the Midwest. Our gratest problem with making the composted landscape mulch was we had absolutely no advertising, no public relations.

I wanted to have an open house and bring a lot of our big customers to our site and show them what we did and asked what we could do for them. I wanted to film our operation and go to like a Home Depot or Menards on the weekends, run the video, and answer questions about the product for more sales. I wanted to rent a large billboard with a wilted flower on the left half of the sign and a healty flower on the right half, and have the slogan,

"Got Compost" in the middle for the months of March through July.

The management was changing and public relations and advertising was not a priority. I concentrated on my job, everything that was excess or we were being paid for got dumped on me and we made beautiful compost out of all of it. There was never an incentive to go out and sell our product other than word of mouth. Fred's grandchildren started to make the calls and it was not going very well. It was 2008, all the track contracts had been lost, all the trucking of that stable bedding had been lost, and a lot of compost customers were being lost because of a lack of public relations. Things will change again in the very near future for me.

My Next Forty Years

There has been a tremendous amount of change in my life, both in the world of composting and in the world of farming. Most of the change, unfortunately, was due to death of someone listed on my most respected list. I lost my dad when he was only fifty nine. I lost both Gorman and Fred when they were very young men. Farming has drastically changed so a single individual can virtually farm a thousand acres by himself. In the fifties, a farmer rasied his entire family on less than two hundred acres tending to that parcel his whole life. Now the machinery is so large and the chemicals and seed are so good, that same two hundred acres only takes a day to plant and two days to harvest out of the whole year. Now the major concern is the amount of money that changes hands between the landowners, the fertilizer companies, and the seed dealers, not to mention the bankers. I seen the writing on the wall in 1986 and quite farming while I still had some money left in my pocket. Drought and low prices have a way of getting to you and letting you know when enough is enough.

Most of the change in composting around the world has come from the results of reducing or preventing odors. Anytime thermophilic bacteria start to eat on carbohydrates and begin to multiply, odors are usually produced. This has gone on since man started to control the process and make money from it. Campbell's learned they could no longer produce compost next to their mushroom houses and moved the process 155 miles away for a time. That got too expensive, and they found Gorman O'Reilly to do the job just forty miles away for a time until he died and his wife and daughter took over. That changed, and they found Fred Noorlag to do the job until he died and his family ably took over for a time.

We thought Compost Product's would be closed after a seven year lawsuit, but we continued to make mushroom compost until Money's went bankrupt and closed the old Prince Crossing farm down. All the spent compost that came from that farm gave us an opening to produce and fill that void when the farm was gone and we did that successfully with the excess stable bedding and a lot of other feedstocks. A lot of changes with the same end product as the result. Gorman used to have a saying that when something bad occurs, "it was like shooting into a covey of quail to see how long it takes them to regroup". My entire list of most respected probably never heard that old saying, but all of them showed that ability to regroup when times got rough.

In our years we ground hay for Campbell's down at the Arsenal, we took the tub grinder to the bales because it was logistically much easier than bringing all those bales to Gorman's home base to do the job. It worked well until the loader we used to fill the tub grinder and load the trucks caught fire and was totally destroyed. The Arsenal was producing bombs about four hundred yards from the incident and their management asked us to never grind in the Arsenal again for safety reasons. Gorman made the decision to move the grinding process to his home base which meant we had to move all those round bales about ten miles to do the job. We added about four feet to the width of three old flat bed semi-trailers so we could haul double the number of bales normally. We bought a new and much better loader to feed the grinder and load trucks. We were able to pull the trailers with farm tractors so no truck drivers were needed, and two weeks later we were grinding hay at Joliet. I would say we quickly regrouped like that covey of quail Gorman always talked about.

I was never aware of the bills coming in or going out at Compost Product's. When Money's went bankrupt, our company was owed more than three hundred thousand for compost delivered to Prince Crossing and stable bedding delivered to LaSalle County to be composted. The logistics of the system worked out very good for the trucking side of our operation. Our farm was located half way between the Princeton mushroom farm and the Chicago racetracks. Trucks leaving Chicago would go to Princeton, unload, and a lot of times bring back spent material to the Chicago market or to me for storage and future delivery. Trucks would also leave our farm with finished compost to West Chicago, unload, and go into the tracks to pick up a load of stable bedding and return to Princeton or me and dump the fresh stable bedding for composting. When Money's went bankrupt, that covey of quail was shot into and it was just a matter of time before we would regroup. The tracks started to pay more for removal because there was less of a demand for the material, and we went out and sought other feedstock's that people were willing to pay to get rid of. It all continued to work extremely well until the grandchildren started to make some of the decisions.

There has been a lot of change as far as how compost is made around the world. The biology has remained the same, but the physical process has under went major changes and it is all related to odor prevention. Fifty years ago there was a lot of smells and a lot of pollution in the air. Now everything has been cleaned up so well that if anybody smells an odor, all hell breaks loose. Operations that produce compost for the mushroom or the landscape market have a lot of new tools to use as long as somebody has enough money to pay for them so all that hell won't break loose.

I have seen mushroom operations that pre-wet the raw feedstocks, add all the nitrogen supplementation, mix completely, and load all this into an enclosed tunnel. The tunnel is then sealed off and air is pumped in from time to time so the temperature is optimally controlled to within a degree or two for best composting. All the odors are sucked out of the tunnel and pumped through a five foot deep bio-filter made of woodchips and finished compost to eliminate all the smells produced from a normal compost process. After a week in the tunnel, the compost is brought out and filled into the mushroom houses for production. All done without a hint of odor and very little machinery, but it all comes with a very hefty price tag.

Some of the most extreme compost odors are produced using bio-solids or sludge from waste treatment plants. This material starts off at a C/N of 6/1 which creates a lot of odors when composted. These new operations usually perform all of the composting in an enclosed building with concrete tunnels that are open to the top. Raw ingredients like straw or woodchips are mixed with the sludge to dry the formula down to about fifty-five percent moisture and a C/N of about 30/1. The tunnel is loaded from the end with the proper mix. A turning device sits on top of the tunnel walls on tracks and moves back and forth to turn the material below. The machine will start at the furthest end and dip its turning device down into the compost at a forty five degree angle. As the turner moves forward, the material is thrown back about twenty feet each time. When it gets to the end, enough space has now been made to allow for the new mix of the day to be added to the tunnel. After twelve to fifteen turns in a thirty day period, the oldest material in the tunnel has now moved to the far end, and is ready to be used and loaded out of the building. All the air from inside the building is again pumped through a bio-filter eliminating all the odors produced. I have seen operations set up within a hundred yards of residents without a single odor complaint ever filed. It is expensive, but the technology is there.

There are operations that use large reusable mesh bags that are fifteen feet or more in diameter and up to four hundred feet long instead of using a building. They have the ability to fill these bags with a mixture of proper moisture and C/N, seal them off, pump air in or out, and run this air again through a bio-filter for the three or four week composting cycle. The material is then removed and piled for curing long after the most odorous part of the process is completed. The bags are reusable and also repel water off the material as good as if it was in a building.

I stated early in the book that composting was a great way to remediate a lot of today's problems. I have seen operations that have taken contaminated soil filled with TNT and compost that material down to its original compounds and safe to use as soil again. I have seen soil filled with Diesel Fuel and Gasoline and with proper composting, all are broken down to the most natural basis compounds and soil can be used for anything.

With the application of compost, vegetation will require less water. Place it around the flowers or any other type of growing plant, and less water will be required. Top dress the lawn or put a layer down before installing sod, and you will only need one third the water to keep everything green. The more organic matter in the soil, the more moisture will be held. In a lot of the areas in California and Florida where expensive crops are grown, compost has reduced the water uasage by fifty percent to produce the same volume of crop where there has been so much talk about water shortages.

Fertilizer could be eliminated with the proper application every year of a well done compost. The best part of compost is that it releases its nutrients very slowly and is not here today and gone tomorrow. Many trace elements will be added to the soil that you will not get with dry fertilizer. There is continuing research and trials that compost has a lot of suppressive traits in reducing harmful insects and disease. My own experiment with corn production with the Gypost showed a big decrease in rootworm destruction and a lot bigger yield. Places that I applied Gypost for soybean production showed a very big decrease in White Mold that was such a big problem in 2009 in our area.

It's hard not to talk about composting in regards to landfill space. If all the organics were removed from the inbound gabage, there would be a sixty percent reduction in volume and extending the life of a landfill for decades. There is still way too much organic material that could be composted going into our landfills. The only thing holding up the removal and eventual composting is the tipping fee at the gate of these landfills. When rates get above eighty dollars per ton to dump will it be profitable for someone to stand at a sorting station and separate all the organic material that is being land filled everyday.

When organics are buried, the results are methane production because of the anaerobic decomposition of all the materials that could be composted in the proper way. The art of collecting methane usually does not begin at a landfill for at least five or six years after methane production has already started and is being released into the atmosphere. Everybody complains about carbon dioxide production in relation to global warming, but methane is thirty times worse for the atmosphere and global warming than carbon dioxide. Most of the eventual recapture of this methane is only about sixty percent effective at the very best at any landfill. If organics were eliminated from the start, there would be little need for methane collection. A lot of local governments talk about seperation and diversion out of one side of their mouths, but the bottom line is a lot of their income to run the government comes from tons delivered to a landfill, and talk about attracting more garbage out of the other side of their mouth.

Another great attribute of composted material is that any of the nutrients in the organic matter will not leach out into a retention pond or stream once it is applied back on the earth where it all started. What is held by the compost will stay with compost until it is absorbed by the plant that needs it. You won't get the green retention ponds from fertilizer runoff or fish kills from a composted product. One of the biggest uses recently has been the use of compost for road reconstruction and hard to grow areas. Once compost is put down, it will not move like black dirt, and the seeding will get a chance to take hold and become established very quickly.

You have probably seen the use of black plastic silt fences around construction sites. Not only are they a waste of money, but most of the time they do not do what they are suppose to do. If a two foot barrier of compost would be used instead, the erosian potential is almost non-existant. There are companies that are making bio-degradable socks that are filled with compost and placed around construction sites in place of these silt fences. Once the project is done, the socks outer layer has degraded and you just flat blade the compost out for plants to use in the future. No waste is produced. Texas is the biggest advocate of compost for erosian control and the use of compost filled socks of any state in the nation. Many states would be wise to follow their lead.

All these new innovations and ideas are the result of a lot of research being done by a lot of people around the world. Compost is becoming a lot more cool than ever before. In an effort to help the environment, a lot more yardwaste, bio-solids, and organic composting and recycling is taking place everywhere. I understand that San Francisco is now at seventy five percent diversion from landfills. That is an unbelievable number from a city that size. For all that to happen, it has to be profitable or it will never get done anywhere.

It is now 2009 and four days before Christmas and change rears its ugly head again. Fred's grandchildren have decided they can continue the operation a lot cheaper without me. I am sure they can do it cheaper, but I don't think they can do it better. Despite it all, I wish them the best. I hope I am able to regroup like that infamous covey of quail Gorman always talked about. It has been difficult not pulling my old work shoes on seven days per week. I won't have to face another miserable winter outside like I have done for the last forty years. I will miss getting my hands dirty, making day to day decisions for the betterment of the company, and solving problems with my crew. I still can't sleep past four in the morning as hard as I try. The toughest part was getting cut off cold turkey from one of my addictions and don't know if my farming addiction will be enough. I'am hoping it will.

I do have a few regrets I must mention before I close out this book. I always put my farming and composting before my family, my health, and my well being. I never knew anything but work since a young kid. It seems the only time I accomplished anything in my mind was when I was working. Nothing else mattered to me. When I was in grade school or high school, I couldn't wait to get home and work in the garden or work for one of the truck farmers in the area. In college, I very rarely stayed any of the weekends at Champaign so I could go home and help Gorman. I would drive back to college dead tired late on Sunday evening and get ready for the week ahead with my classes. My wife and I would have never met if it was not for a wet spell close to the fourth of July in 1973 and was unable to work for a couple of days. Looking back forty plus years, I can say one thing for sure, work is highly overrated.

After my bout with rheumatic fever, I went right back working like nothing had happened despite what the doctors said or my mother's wishes. I had all the makings of a good athlete, but was denied being one because of insurance. When my chance came to manage the composting at Gorman's and at Compost Product's, I went at it head on without any respect to my health or well being. I don't know a lot of people who have worked in seventy-five below wind chill and actually accomplished something and for what? We seem to have done it every winter and now looking back, what did it actually get me except gray hair and many years on my body that no one can replace.

Many will work for the money or the praise. In all my forty years of composting and farming, there was plenty of money, but never any praise. You did good the last time, lets make the next time even better. If you never heard anything about anything, that was the extent of the praise. The biggest thing that has kept me going was the outright addiction to both composting and farming from my earliest years. I was compost when compost wasn't cool. Could not live without it, could hardly live with it.

Acknowledgments; My List Of Most Respected

I will never forget all my Mom and Dad for me in those early years. I learned about good work ethic, about how to grow something and sell it, and the pleasure of growing up in a rural environment. I just wish Mom and Dad were here to see the success I achieved as I got older. I want to mention my Grandmother and Grandfather on my Moms side. They never lived beyond their means. They subsisted for many years with very little but always felt they had a lot. Many people could learn a lot from these two and how to live a fuller life.

I don't have enough words to describe the support, encouragement, and financial help I received from Ron Deininger, my high school agriculture teacher. Those four years in high school and in FFA were remarkable looking back forty years. It was because of him I got the chance to attend the University of Illinois, and could not have done that without him.

You probably noticed a large portion of the book is devoted to my first full time employer Gorman O'Reilly. He was a wonderful person to know and to work for over seventeen years early in my life. He kind of took over and looked after me after my father died in 1972. He was always the innovator, he was always ahead of everybody else in the farm community, he became the largest operator in the area, and I was glad to be a part of that for so long. He was able to feed my addiction to both Farming and Composting. We did some very amazing things in those seventeen years before he died very suddenly in 1982. I sure miss him. When ever I get into some difficult position, I just wonder what Gorman would do and everything usually works out very well. There is not a day that goes by that I don't think about him and never got the chance to thank him for all he did for me.

My list of most respected includes Jim Frezzo and his family, Lou Pia and his family, Bob Pannell and his family, and want to thank them for all they did those first few years to help me become successful with composting. They sure made the job of mushroom composting a lot easier after my first visit to the mushroom capital of the world in 1973.

Fred Noorlag gave me a second chance to feed my addiction to compost and farming. I was able to continue to farm my land until 1988, then I started to farm his land with the hay adventure until about 1996. He let me run Compost Product's like it was mine. Another of the most respected I never got the chance to thank for all he did for me. His family took over after his death and the transition was seemless two different times. We had to change and adapt after Fred died, and after Money's went bankrupt. I want to thank his family for the twenty plus years of allowing me to run their operation for them.

I want to mention Dr. Bob Miller, Ramon Barajas, Bill Edholm, Jim Elliot, Tom Gannon, and Ramon Nieto that insured in me the confidence to make compost for Campbell Soup for twenty plus years. I never realized while I was doing the composting process for Prince Crossing that there was over 250 people relying on me and my crew to do the job no matter what the weather was or the problems that arose. I get a little nervous now just thinking about that and for all those years. These five men and about a hundred others in the mushroom division all made an impact on my life and want to say thank you to all of them.

I had the great fortune to meet Mr. Gary Vermeer from Pella, Iowa. He impressed me back in 1969, and he and his company are still impressing me today with all their inventions and well built and designed equipment that they sell.

During my years of farming, Jack Tyler, Don Borchardt, and Leo Bernhard all had a tremendous effect on my career and will be forever grateful. Because I worked for Gorman, I was able to meet L. Eugene Smith and Gene Shoup. Both were enjoyable to meet and learn about what a man can do with his talent. I was on the cutting edge of what most farmers now take for granted with the invention of large corn planters.

Tom Creech has had a big impact on my life and have known him for a good portion of my forty year addiction to composting and stable bedding, or as he calls it, muck. It has been a pleasure to work with him and his entire team in Lexington, Kentucky. He has always has a lot of ideas about a lot of things that have become very successful. Maybe someday, I will actually go and work for him if the chance is there.

When we transitioned from mushroom compost to composted landscape mulch, Les Kulhman and his wife Joan from Sterling, Colorado supplied us with our first windrow turner and a lot of support to get started. The K/W turner sure made the change a lot easier than I had envisioned. Sharon Barnes from Huron, Ohio gave me a lot of confidence to make the change from mushroom compost to landscape mulch and want to thank her for all of the support for me and the entire compost industry. Also want to say thanks to Tim O'Hara of Wildcat Manufacturing in Freeman, South Dakota for all of his support in the world of windrow composting. If I ever become rich, I will go to the Freeman area and buy some of the most beautiful land I have ever seen. There was a sign along the road I saw when I made the trip up to see our machine before delivery. It read, "speed limit 75 or what ever feels comfortable". It was quite a place.

It sure has been a wild ride for my wife and I. There is no way to ever thank her for all the support over the last thirty six years of my forty year addiction to composting and farming. I was able to farm and compost my whole life and would not change a thing except the brutal weather in Chicago every Winter. I wish a lot of my mentors were still here with us today and wonder what it would be like to have them here with us? Change is inevitable and I have never been afraid of change. I never reached my level of incompetence with anything I ever tackled. I thought that I regrouped very quickly like that covey of quail Gorman always talked about. I will never forget that "I Was Compost When Compost Wasn't Cool", and those people are far and few between.

LaVergne, TN USA
18 November 2010
205392LV00004B/4/P
9 781452 049229